TEN ACRES ENOUGH

A PRACTICAL EXPERIENCE,

SHOWING

**HOW A VERY SMALL FARM MAY BE MADE
TO KEEP A VERY LARGE FAMILY.**

TEN ACRES ENOUGH

A PRACTICAL EXPERIENCE,

SHOWING

HOW A VERY SMALL FARM MAY BE MADE
TO KEEP A VERY LARGE FAMILY.

WITH

Extensive and profitable Experience

IN

THE CULTIVATION OF THE SMALLER FRUITS.

NEW YORK:
PUBLISHED BY JAMES MILLER,
(successor to c. s. francis & co.,)
522 BROADWAY.
MDOOOLXIV.

ISBN: 978-1-6673-0637-7 paperback
ISBN: 978-1-6673-0638-4 hardcover

PREFACE.

THE man who feeds his cattle on a thousand hills may possibly see the title of this little volume paraded through the newspapers; but the chances are that he will never think it worth while to look into the volume itself. The owner of a hundred acres will scarcely step out of his way to purchase or to borrow it, while the lord of every smaller farm will be sure it is not intended for him. Few persons belonging to these several classes have been educated to believe Ten Acres Enough. Born to greater ambition, they have aimed higher and grasped at more, sometimes wisely, sometimes not. Many of these are now owning or cultivating more land than their heads or purses enable them to manage properly. Had their ambition been moderate, and their ideas more practical, their labor would be better rewarded, and this book, without doubt, would have found more readers.

The mistaken ambition for owning twice as much land as one can thoroughly manure or profitably cultivate, is the great agricultural sin of this country. Those who commit it, by beginning wrong, too frequently continue wrong. Owning many acres is the sole idea. High cultivation of a small tract, is one of which they have little knowledge. Too many in these several classes think they know enough. They measure a man's knowledge by the number of his acres. Hence, in their eyes the owner of a plot so humble as mine must know so little as to be unable to teach them any thing new.

Happily, it is not for these that I write, and hence it would be unreasonable to expect them to become readers. I write more particularly for those who have not been brought up as farmers—for that numerous body of patient toilers in city, town, and village, who, like myself, have struggled on from year to year, anxious to break away from the bondage of the desk, the counter, or the workshop, to realize in the country even a moderate income, so that it be a sure one. Many such are constantly looking round in this direction for something which, with less men- tal toil and anxiety, will provide a maintenance for a growing family, and afford a refuge for advancing age— some safe and quiet harbor, sheltered from the constantly recurring monetary and political convulsions which in this country so suddenly reduce men to poverty. But these inquirers find no experienced pioneers to lead the way, and they turn back upon themselves, too fearful to go forward alone. Books of personal experience like this are rare. This is written for the information of the class referred to, for men not only willing, but anxious to learn. Once in the same predicament myself, I know their longings, their deficiencies, and the steps they ought to take. Hence, in seeking to make myself fully understood, some may think that I have been unnecessarily minute. But in setting forth my own crudities, I do but save others from repeating them. Yet with all this amplification, my little contribution will occasion no crowding even upon a book-shelf which may be already filled.

I am too new a farmer to be the originator of all the ideas which are here set forth. Some, which seemed to be appropriate to the topic in hand, have been incorporated with the argument as it progressed; while in some instances, even the language of writers, whose names were unknown to me, has also been adopted.

CONTENTS

TEN
ACRES
ENOUGH

CHAPTER I.
CITY EXPERIWENCES –
MODERATE EXPECTATIONS.

My life, up to the age of forty, had been spent in my native city of Philadelphia. Like thousands of others before me, I began the world without a dollar, and with a very few friends in a condition to assist me. Having saved a few hundred dollars by dint of close application to business, and avoiding taverns, oyster-houses, theatres, and fashionable tailors, I married and went into business the same year. These two contemporaneous drafts upon my little capital proving heavier than I expected, they soon used it up, leaving me thereafter greatly straitened for means. It is true my business kept me, but as it was constantly expanding, and was of such a nature that a large proportion of my annual gain was necessarily invested in tools, fixtures, and machinery, I was nearly always short of ready cash to carry on my operations with comfort. At certain times, also, it ceased to be profitable. The crisis of 1837 nearly ruined me, and I was kept struggling along during the five succeeding years of hard times, until the revival of 1842 came round. Previous to this crisis, necessity had driven me to the banks for discounts, one of the sore evils of doing business upon insufficient capital. As is always the case with these institutions, they compelled me to return the borrowed money at the very time it was least convenient for me to do so – they needed it as urgently as myself. But to refund them I was compelled to borrow elsewhere, and that too at excessive rates of interest, thus increasing the burden while laboring to shake it off.

Thousands have gone through the same unhappy experience, and been crushed by the load. Such can anticipate my trials and privations. Yet I was not insolvent. My property had cost me far more than I owed, yet if offered for sale at a time when the whole community seemed to want money only, no one could have been found to give cost. I could not use it as the basis of a loan, neither could I part with it without abandoning my business. Hence I struggled on through that exhausting crisis, haunted by perpetual fears of being dishonored at bank, – lying down at night, not to peaceful slumber, but to dream of fresh expedients to preserve my credit for to-morrow. I have sometimes thought that the pecuniary cares of that struggle were severe enough to have shortened my life, had they been much longer protracted.

Besides the mental anxieties they occasioned, they compelled a pinching economy in my family. But in this latter effort I discovered my wife to be a jewel of priceless value, coming up heroically to the task, and relieving me of a world of care. Without her aid, her skill, her management, her uncomplaining cheerfulness, her sympathy in struggles so inadequately rewarded as mine were, I should have sunk into utter bankruptcy. Her economy was not the mean, penny-wise, pound foolish policy which many mistake for true economy. It was the art of calculation joined to the habit of order, and the power of proportioning our wishes to the means of gratifying them. The little pilfering temper of a wife is despicable and odious to every man of sense; but there is a judicious, graceful economy, which has no connection with an avaricious temper, and which, as it depends upon the understanding, can be expected only from cultivated minds. Women who have been well educated, far from despising domestic duties, will hold them in high respect, because they will see that the whole happiness of life is made up of the happiness of each particular day and hour, and that much of the enjoyment of these must depend upon the punctual practice of virtues which are more valuable than splendid.

If I survived that crisis, it was owing to my wife's. admirable management of my household expenses. She saw that our embarrassment was due to no imprudence or neglect of mine. She thus consented to severe privations, uttering no complaint, hinting no reproach, never disheartened, and so rarely out of humor that she never failed to welcome my return with a smile.

But in this country one convulsion follows another with disheartening frequency. I lived through that of 1837, paid my debts, and bad managed to save some money. My wife's system of economy had been so long adhered to, that in the end it became to some extent habitual to her, and she still continued to practise great frugality. I became insensibly accustomed to it myself. Children were multiplying around us, and we thought the skies had brightened for all future time. When in difficulty, we had often debated the propriety of quitting the city and its terrible business trials, and settling on a few acres in the country, where we could raise our own food, and spend the remainder of our days in cultivating ground which would be sure to yield us at least a respectable subsistence. We had no longing for excessive wealth: a mere competency, though earned by daily toil, so that it was reasonably sure, and free from the drag of continued indebtedness to others, was all we coveted.

I had always loved the country, but my wife preferred the city. I could take no step but such as would be likely to promote her happiness. So long as times continued fair, we ceased to canvass the propriety of a removal. We had children to educate, and to her the city seemed the best and most convenient place for qualifying them for future usefulness. Then, most of our relations resided near us. Our habits were eminently social. We had made numerous friends, and among our neighbors there had turned up many valuable families. We felt even the thought of breaking away from all these cordial ties to be a trying one. But the refuge of a removal to the country had taken strong hold of my mind.

Indeed, it may be said that I was born with a passion for living on a farm. It was fixed and strengthened by my long experience of the business vicissitudes of city life. For many years I had been a constant subscriber for several agricultural journals, whose contents I read as carefully as I did those of the daily papers. My wife also, being a great reader, came in time to study them almost as attentively. Every thing I saw in them only tended to confirm my longing for the country, while they gave definite views of what kind of farming I was fit for. In fact they educated me for the position before I assumed it. I am sure they exercised a powerful influence in removing most of my wife's objections to living in the country. I studied their contents as carefully ^s did the writers who prepared them. I watched the reports of crops, of experiments, and of profits. The leading idea in my mind was this – that a man of ordinary industry and intelligence, by choosing a proper location within hourly reach of a great city market, could so cultivate a few acres as to insure a maintenance for his family, free from the ruinous vibrations of trade or commerce in the metropolis. All my reading served to convince me of its soundness. I did not assume that he could get rich on the few acres which I ever expected to own; but I felt assured that he could place himself above want. I knew that his peace of mind would be sure. With me this was dearer than all. My reading had satisfied me that such a man would find Ten Acres Enough, and these I could certainly command.

As I did not contemplate undertaking the management of a large grain farm, so my studies did not run in that direction. Yet I read every thing that came before me in relation to such, and not without profit. But I graduated my views to my means, and so noted with the utmost care the experiences of the small cultivators who farmed five to ten acres thoroughly. I noted their failures as watchfully as their successes, knowing that the former were to be avoided, as the latter were to be imitated. As opportunity offered, I made repeated excursions, year after year, in every direction around Philadelphia, visiting the small farmers or truckers who supplied the city market with fruit and vegetables, examining, inquiring, and treasuring up all that I saw and heard. The fund of knowledge thus acquired was not only prodigious, but it has been of lasting value to me in my subsequent operations. I found multitudes of truckers who were raising large families on five acres of ground, while others, owning only thirty acres, had become rich.

On most of these numerous excursions I was careful to have my wife with me. I wanted her to see and hear for herself, and by convincing her judgment, to overcome her evidently diminishing reluctance to leaving the city. My uniform consideration for her comfort at last secured the object I had in view. She saw so many homes in which a quiet abundance was found, so many contented men and women, so many robust and bouncing children, that long before I was ready to leave the city, she was quite impatient to be gone.

CHAPTER II.
PRACTICAL VIEWS SAFETY OF INVESTMENTS IN LAND.

THERE was not a particle of romance in my aspirations for a farm, neither had I formed a single visionary theory which was there to be tested. My notions were all sober and prosaic. I had struggled all my life for dollars, because abundance of them produces pecuniary comfort: and the change to country life was to be, in reality, a mere continuation of the struggle, but lightened by the assurance that if the dollars thus to be acquired were fewer in number, the certainty of earning enough of them was likely to be greater. Crops might fail under skies at one time too watery, at another too brassy, but no such disaster could equal those to which commercial pursuits are uninterruptedly exposed. They have brassy skies above them as well as farmers. For nearly twenty years I had been hampered with having notes of my own or of other parties to pay; but of all the farmers I had visited, only one had ever given a note, and he had made a vow never to give another. My wife was shrewd enough to observe and remark on this fact at the time, it was so different from my own experience. She admitted there must be some satisfaction in carrying on a business which did not require the giving of notes.

Looking at the matter of removal to the country in a practical light, I found that in the city I was paying three hundred dollars per annum rent for a dwelling-house. It was the interest of five thousand dollars; yet it afforded nothing but a shelter for my family. I might continue to pay that rent for fifty years, without, at the end of that time, having acquired the ownership of either a stone upon the chimney, or a shingle in the roof. If the house rose in value, the rise would be to the owner's benefit, not to mine. It would really be injurious to me, as the rise would lead him to demand an increase of his rent. But put the value of the house into a farm, or even the half of it – the farm would have a dwelling- house upon it, in which my family would find as good a shelter, while the land, if cultivated as industriously as I had always cultivated business, would belie the flood of evidence I had been studying for many years, if it failed to yield to my efforts the returns which it was manifestly returning to others. We could live contentedly on a thousand dollars a year, and here we should have no landlord to pay. My wife, in pinching times, has financiered us through the year on several hundred less. I confess to having lived as well on the diminished rations as I wanted to. Indeed, until one tries it for himself, it is incredible what dignity there is in an old hat, what virtue in a time-worn coat, and how savory the dinner-table can be made without sirloin steaks or cranberry tarts.

Thus, let it be remembered, my views and aspirations had no tinge of extravagance. My rule was moderation. The tortures of a city struggle without capital, had sobered me down to being contented with a hare competency. I might fail in some particulars at the outset, from ignorance, hut I was in the prime of life, strong, active, industrious, and tractable, and what I did not know I could soon learn from others, for farmers have no secrets. Then I had seen too much of the uncertainty of banks and stocks, and ledger accounts, and promissory notes, to be willing to invest anything in either as a permanency. At best they are fluctuating and uncertain, up to-day and down to-morrow. My great preference had always been for land.

In looking around among my wide circle of city acquaintances, especially among the older families, I could not fail to notice that most of them had grown rich by the ownership of land. More than once had I seen the values of all city property, improved and unimproved, apparently disappear; – lots without purchasers, and houses without tenants, the community so poor and panic-stricken that real estate became the merest drug. Yesterday the collapse was caused by the destruction of the National Bank; to-day it is the Tariff. Sheriffs played havoc with houses and lands incumbered by mortgages, and lawyers fattened on the rich harvest of fees inaugurated by a Bankrupt Law. But those who, undismayed by the wreck around them, courageously held on to land, came through in safety. The storm, having run its course and exhausted its wrath, gave place to skies commercially serene, and real estate swung back with an irrepressible momentum to its former value, only to keep on advancing to one even greater.

I became convinced that safety lay in the ownership of land. In all my inquiries both before leaving the city, as well as since, I rarely heard of a farmer becoming insolvent. When I did, and was careful to ascertain the cause, it turned out that he had either begun in debt, and was thus hampered at the beginning, or had made bad bargains in speculations outside of his calling, or wasted his means in riotous living, or had in some way utterly neglected his business. If not made rich by heavy crops, I could find none who had been made poor by bad ones.

The reader may look back over every monetary convulsion he may be able to remember, and he will find that in all of them the agricultural community came through with less disaster than any other interest. Wheat grows and com ripens though all the banks in the world may break, for seed-time and harvest is one of the divine promises to man, never to be broken, because of its divine origin. They grew and ripened before banks were invented, and will continue to do so when banks and railroad bonds shall have become obsolete.

Moreover, the earthly fund for whose acquisition we are all striving, we naturally desire to make a permanent one. As we have worked for it, so we trust that it will work for us and our children. Its value, whatever that may be, depends on its perpetuity – the continuance of its existence. A man seeks to earn what will support and serve not only himself, but his posterity. He would naturally desire to have the estate descend to children and grandchildren. This is one great object of his toil. What, then, is the safest fund in which to invest, in this country? What is the only fund which the experience of the last fifty years has shown, with very few exceptions, would be absolutely safe as a 'provision for heirs? How many men, within that period, assuming to act as trustees for estates, have kept the trust fund invested in stocks, and when distributing the principal among the heirs, have found that most of it had vanished ! Under corporate insolvency it had melted into air. No prudent man, accepting such a trust, and guaranteeing its integrity, would invest the fund in stocks.

Our country is filled with pecuniary wrecks from causes like this. Thousands trust themselves during their lifetime, to manage this description of property, confident of their caution and sagacity. With close watching and good luck, they may be equal to the task; but the question still occurs as to the probable duration of such a fund in families. What is its safety when invested in the current stocks of the country? and next, what is its safety in the hands of heirs? There are no statistics showing the probable continuance of estates in land in families, and of estates composed of personal property, such as stocks. But every bank cashier will testify to one remarkable fact – that an heir no sooner inherits stock in the bank than the first thing he generally does is to sell and transfer it, and that such sale is most frequently the first notice given of the holder's death.

This preference for investment in real estate will doubtless be objected to by the young and dashing business man. But lands, or a fund secured by real estate, is unquestionably not only the highest security, but in the hands of heirs it is the only one likely to survive a single generation. Hence the wisdom of the common law, which neither permits the guardian to sell the lands of his ward, nor even the court, in its discretion, to grant authority for their sale, except upon sufficient grounds shown, – as a necessity for raising a fund for the support and education of the ward. Even a lord chancellor can only touch so sacred a fund for this or similar reasons. The common law is wise on this subject, as on most others. It is thus the experience and observation of mankind that such a fund is the safest, and hence the provisions of the law.

Those, therefore, who acquire personal property, acquire only what will last about a generation, longer or shorter. Such property is quickly converted into money – it perishes and is gone. But land is hedged round with numerous guards which protect it from hasty spoliation. It is not so easily transferred; it is not so secretly transferred; the law enjoins deliberate formalities before it can be alienated, and often the consent of various parties is necessary. When all other guards give way, early memories of parental attachment to these ancestral acres, or tender reminiscences of childhood, will come in to stay the spoliation of the homestead, and make even the prodigal pause before giving up this portion of his inheritance.

Throughout Europe a passion to become the owner of land is universal, while the difficulty of gratifying it is infinitely greater than with us. It is there enormously dear; here it is absurdly cheap. It is from this universal passion that the vast annual immigration to this country derives its mighty impulse. As it reaches our shores it spreads itself over the country in search of cheap land. Many of the most flourishing Western States have been built up by the astonishing influx of immigrants. In England, every landowner is prompt to secure every freehold near him, be it large or small, as it comes into market. Hence the number of freeholders in that country is annually diminishing by this process of absorption. This European passion for acquiring land is strangely contrasted with the American passion for parting with it.

CHAPTER III.
RESOLVED TO GO – ESCAPE FROM BUSINESS – CHOOSING A LOCATION.

THE last thirty years have been prolific of great pecuniary convulsions. I need not recapitulate them here, as too many of them are yet dark spots on the memory of some who will read this. Their frequency, as well as their recurrence at shorter intervals than at the beginning of the century, are among their most remarkable features, baffling the calculations of older heads, and confounding those of younger ones. As the century advanced, these convulsions increased in number and violence. The whole business horizon seemed full of coming storms, which burst successively with desolating severity, not only on merchants and manufacturers, but on others who had long before retired from business. No one could foresee this state of things. I will not stop to argue causes, but confine myself to facts which none will care to contradict.

These disasters made beggars of thousands in every branch of business, and spread discouragement over every community. I passed through several of them, striving and struggling, and oppressed beyond all power of description. How many more the community was to encounter I did not know; but I conceived it the part of prudence to place myself beyond the circle of their influence before I also had been prostrated.

In spite of the losses thus encountered, I had been saving something annually for several years, when the stricture of 1854 came on, premonitory of the tremendous crash of 1857. Most unfortunately for my comfort, that stricture seemed to fall with peculiar severity on a class of dealers largely indebted to me. Many of them became embarrassed, and failed to pay me at the time, while to this day some of them are still my debtors. My old experiences of raising money revived, and to some extent I was compelled to go through the humiliations of similar periods. But the stricture was of brief duration, and I closed the year in far better condition than I had anticipated.

But the trials of that incipient crisis determined me to abandon the city. I found that by realizing all I then possessed, I could command means enough to purchase ten to twenty acres, and I had grown nervous and apprehensive of the future. While possessed of a little, I resolved to make that little sure by investing it in land. I had worked for the landlord long enough. My excellent wife was now entirely willing to make the change, and our six children clapped their hands with joy when they heard that "father was going to live in the country."

I had long determined in my mind what sort of farming was likely to prove profitable enough to keep us with comfort, and that was the raising of small fruits for the city markets. My attention had always been particularly directed to the berries. Some strawberries I had raised in my city garden with prodigious success. My friends, when they heard of my project, expressed fears that the market would soon be glutted, not exactly by the crops which I was to raise, but they could not exactly answer how. They confessed that they were extremely fond of berries, and that at no time in the season could they afford to eat enough; a confession which seemed to explode all apprehension of the market being overstocked.

But my wife and myself had both examined the hucksters who called at the door with small fruits, as to the monstrous prices they demanded, and had begged them, if ever a glut occurred, that they would call and let us know. But none had ever called with such information. It was the same thing with those who occupied stalls in the various city markets. They rarely had a surplus left unsold, and their prices were always high. A glut of fruit was a thing almost unknown to them. It was a safe presumption that the market would not be depressed by the quantity that I might raise.

But here let me say something by way of parenthesis, touching this common idea of the danger of overstocking the fruit-market of the great cities. It is a curious fact that this idea is entertained only by those who are not fruit-growers. The latter never harbored it. Their whole experience runs the other way, they know it to be a gross absurdity. Yet, somehow, the question of a glut has always been debated. Twenty years ago the nurserymen were advised to close up their sales and abandon the business, as they would soon have no customers for trees – everybody was supplied. But trees have continued to be planted from that day to this, and where hundreds were sold twenty years ago, thousands are disposed of now. Old-established nurseries have been trebled in size, while countless new ones have been planted. The nursery business has grown to a magnitude truly gigantic, because the market for fruit has been annually growing larger, and no business enlarges itself unless it is proved to be profitable.

The market cannot be glutted with good fruit. The multiplication of mouths to consume it is far more rapid than the increase of any supply that growers can effect. Within ten years the masses have had a slight taste of choice fruits, and but little more. Indulgence has only served to whet their appetites. The more of them there is offered in the market, the more will there be consumed. Every huckster in her shamble, every vender of peanuts in the street, will testify to this. The modern art of semi-cookery for fruit, and of preserving it in cans and jars, has made sale for enormous quantities of those choicer kinds which return the highest profit to the grower. It is in the grain-market that panic often rages, but never in the fruit-market. If it ever enters the latter, the struggle is to obtain the fruit, not to get rid of it.

The proper choice of a location was now to be the great question of my future success. I had determined on giving my attention to the raising of the smaller fruits for the great markets of New York and Philadelphia. I must therefore be somewhere on or near the railroad between those cities, and as near as possible to a station. The soil of Pennsylvania, near Philadelphia, was too heavy for some of the lighte fruits. New Jersey, with its admirable sandy loam light, warm, and of surprisingly easy tillage, was proverbially adapted for the growth of all market produce, whether fruit or vegetable, and was at the same time a week or two earlier. Land was far cheaper, there was no State debt, taxes were merely nominal, and an acre that could be bought for thirty dollars could be made four times as productive as an acre of the best wheat land in Pennsylvania. Such results are regularly realized by hundreds of Jerseymen from year to year.

It was also of easy access from the city for manure-boats. Every town within the range of my wants was well supplied with churches, schools, and stores, together with an intelligent and moral population. I should be surrounded by desirable neighbors, while an hour's ride by steamboat or railroad would place me, many times daily, among all my ancient friends in the city. We should by no means become hermits. I knew the country so well from my numerous visits among the fruit-growers, when in search of information, as to anticipate but little difficulty in finding the proper location.

By the mere accident of a slight revival in business in the early part of 1855, a party came along who was thus induced to purchase my stock and machinery. Luckily, he was able to pay down the whole amount in cash. I received what I considered at the time an excellent price; but when I came to settle up my accounts and pay what I owed, I found, to my extreme disappointment, that hut a little over two thousand dollars remained.

This sum was the net gain of many years of most laborious toil. Was it possible for farming to be a worse business than this? I had made ten times as much, but my losses had been terrible. This, with my personal credit, was all the surplus I had saved. I remember now, that when thus discovering myself to be worth so little, I half regretted having given up my business for what then appeared to me so inadequate a sum. When selling, I was jubilant and thankful – when settled up, I was full of regrets. I ought to have had more. So difficult is it for the human mind to be satisfied with that which is really best.

But I little knew what the future was to bring forth, and how soon my want of thankfulness was to be changed into the profoundest conviction that I had providentially escaped from total ruin, and come out comparatively rich. I had made myself snug upon my little farm when the tornado of 1857 toppled my former establishment into utter ruin. My successor was made a bankrupt, and his business was destroyed, leaving him overwhelmed with debt. He had lost all, while I had saved all. Had I not sold when I did, and secured what the sale yielded me, I too should have been among the wrecks of that terrific visitation.

But I heard its warring in the quiet of my little farm-house, where it brought me neither anxiety nor loss. My position was like that of one sitting peacefully by his wintry fireside, gazing on the thick storm without, and listening to the patter of the snow-flakes as the tempest drove them angrily against the window-pane, while all within was calm and genial. Instead of regrets for what I had failed to grasp, my heart overflowed with thankfulness for the comparative abundance that remained to me. My peace of mind was perfect. The unspeakable satisfaction was felt of being out of business, out of debt, out of danger – not rich, but possessed of enough. The thoughtful reader may well believe that subsequent disturbances, rebellion, war, and even a more wide-spread bankruptcy – from all which my humble position made me secure – have only served to intensify my gratitude to that Divine Providence which so mercifully shaped my ways.

CHAPTER IV.
BUYING A FARM – A LONG SEARCH –
ANXIETY TO SELL – FORCED TO QUIT.

As already stated, I had in round numbers a clear two thousand dollars, with which to buy and stock a farm, and keep my family while my first crops were growing. As I was entirely free from debt, so I determined to avoid it in the future. Debt had been the bitter portion of my life, not from choice, but of necessity. My wife took strong ground in support of this resolution – what we had she wanted us to keep. I had too long been aided by her admirable counsel to reject it now. She had a singular longing for seeing me my own landlord. Her resolution was a powerful strengthener of my own convictions.

Thus resolved, we set out in the early part of March to seek a home. I was particular to take my wife with me – I wanted her to aid in choosing it. She was to occupy it as well as myself. She knew exactly what we wanted as regarded the dwelling-house, – the land department she left entirely to my judgment. I was determined that she should be made comfortable from the start, not only because she deserved to be made so, but to make sure that no cause for future discontent should arise. Indeed, she was really the best judge in this matter. She knew what the six children needed; she was the model of a housekeeper; there were certain little conveniences indispensable to domestic comfort to be secured, of which she knew more than I did, while her judgment on most things was so correct, that I felt confident if she were fully satisfied, the whole enterprise would be a successful one.

I loved her with the fervor of early married life – she had consented to my plans – she was willing to share whatever inconveniences might belong to our new position – was able to lighten them by her unflagging cheerfulness and thrift – and I was unwilling to take a single step in opposition either to her wishes or her judgment. Indeed, I had long since made up my mind, from observation of the good or bad luck of other men, that he who happens to be blessed with a wife possessing good sense and good judgment, succeeds or fails in life according as he is accustomed to consult her in his business enterprises. There is a world of caution, shrewdness, and latent wisdom in such women, which their husbands too frequently disregard to their ruin.

I am thus particular as to all my experiences; for this is really a domestic story, intended for the multitudes who have suffered half a lifetime from trials similar to mine, and who yet feel ungratified longings for some avenue of escape. My object being to point out that through which I emerged from such a life to one of certainty and comfort, the detail ought to be valuable, even if it fail to be interesting. It is possible that I may sink the practical in the enthusiastic, and prove myself to be unduly enamored of my choice. But as. it is success that makes the hero, so let my experience be accepted as the test.

I had settled it in my mind that I would use a thousand dollars in the purchase of land, and that I could make Ten Acres Enough. This I was determined to pay for at once, and have it covered by no man's parchment. But when we set out on our search, we found some difficulties. Every county in New Jersey contained a hundred farms that were for sale. Most of them were too large for my slender purse, though otherwise most eligibly situated. Then we must have a decent house, even if we were forced to put up with less land. Numerous locations of this kind were offered. The trouble was – keeping my slender purse in view – that the farms were either too large or too small. My wife was not fastidious about having a fine house. On the contrary, I was often surprised to find her pleased with such as to me looked small and mean. Indeed, it seemed, after ten days' search, that the tables had been turned – she was more easily suited than myself. But the same deference which I paid to her wishes, she uniformly paid to mine.

It was curious to note the anxiety of so many landowners to sell, and to hear the discordant reasons which they gave for desiring to do so. The quantity in market was enormous. All the real-estate agents had large books filled with descriptions of farms an fancy country-seats for sale, some to be had by paying one-fourth of the purchase-money down, and some which the owners would exchange for merchandise, or traps, or houses in the city. Many of them appeared simply to want something else for what they already had. They were tired of holding, and desired a change of some kind, better if they could make it, and worse if they could not. City merchants, or thriving mechanics, had built country cottages, and then wearied of them – it was found inconvenient to be going to and fro – in fact, they had soon discovered that the city alone was their place. Many such told us that their wives did not like the country.

Others had bought farms and spent great sums in improving them, only to sell at a loss. Farming did not pay an owner who lived away off in the city. Another class had taken land for debt, and wanted to realize. They expected to lose anyhow, and would sell cheap. Then there was another body of owners who, though born and raised upon the land, were tired of country life, and wanted to sell and embark in business in the city. Some few were desirous of going to the West. Change of some kind seemed to be the general craving. As I discovered that much of all this land was covered with mortgages of greater or less amount, it was natural to suppose the sheriff would occasionally turn up, and so it really was. There were columns in some of the county papers filled with his advertisements. I sometimes thought the whole country was for sale.

But yet there was a vast body of owners, many of them descendants of the early settlers, whom no consideration of price could tempt to abandon their inheritances. They seemed to know and understand the value of their ancestral acres. We met with other parties, recent purchasers, who had bought for a permanency, and who could not be induced to sell. In short, there seemed to be two constantly flowing streams of people – one tending from city to country, the other from country to city. Doubtless it is the same way with all our large cities. I think the latter stream was the larger. If it were not so, our cities could not grow in population at a rate so much more rapid than the country. At numerous farm-houses inquiries were made if we knew of any openings in the city in which boys and young men could be placed. The city was evidently the coveted goal with too large a number.

This glut of the land-market did not discourage us. We could not be induced to believe that land had no value because so many were anxious to dispose of it. We saw that it did not suit those who held it, and knew that it would suit us. But we could not but lament over the infatuation of many owners, who we felt certain would be ruined by turning their wide acres into money, and exposing it to the hazards of an untried business in the city. I doubt not that many of the very parties we then encountered have, long before this, realized the sad fate we feared, and learned too late that lands are better than merchandise.

One morning, about the middle of March, we found the very spot we had been seeking. It lay upon the Amboy Railroad, within a few miles of Philadelphia, within gunshot of a railroad station, and on the outskirts of a town containing churches, schools, and stores, with quite an educated society. The grounds comprised eleven acres, and the dwellinghouse was quite large enough for my family. It struck the fancy of my wife the moment we came up to it; and when she had gone over the house, looked into the kitchen, explored the cellar, and walked round the garden, she expressed the strongest desire to make it our home.

There was barn enough to accommodate a horse and cow, with a ton or two of hay, quite an extensive shed, and I noticed that the barnyard contained a good pile of manure which was to go with the property. The buildings were of modern date, the fences were good, and there was evidence that a former occupant had exercised a taste for fruit and ornamental trees, while the garden was in very fair condition. But the land had been wholly neglected. All outside of the garden was a perfect scarecrow of tall weeds, thousands of which stood clear up to the fence top, making sure that they had scattered seeds enough for twenty future crops.

But I noticed that the land directly opposite was in the most admirable condition, and I saw at a glance that the soil must be adapted to the very purpose to which it was to be applied. The opposite ground was matted with a luxuriant growth of strawberries, while rows of stalwart raspberries held up their vigorous canes in testimony of the goodness of the soil. A fine peach-orchard on the same neighboring property, seemed impatient to put forth and blossom unto harvest. The eleven acres could be no worse land than this, and though I had a horror of weeds, yet I was not to be frightened by them. I knew that weeds were more indigenous to New Jersey than even watermelons.

This miniature plantation of eleven acres belonged to a merchant in the city. He had taken it to secure a debt of eleven hundred dollars, but had pledged himself to pay the former owner whatever excess over that sum he might obtain for it. But pledges of that loose character seldom amount to much – the creditor consults his own interest, not that of the debtor. The latter had long been trying to sell, but in vain; and now the former had become equally embarrassed, and needed money even more urgently than the debtor had done. The whole property had cost the debtor eighteen hundred dollars. His views in founding it were similar to mine. He meant to establish for himself a home, to which at some future period he might retire. But he made the sad mistake of continuing in business in the city, and one disaster succeeding another, he had been compelled to abandon his anticipated refuge nearly a year before we came along.

All these facts I learned before beginning to negotiate for the purchase. As the banished man related them to me, going largely into the history of his hopes, his trials, his disappointments, I found cause for renewed thankfulness over my superior condition. With a single exception, his experience had been the counterpart of my own – he had lost all and was loaded with debt, while I had saved something and owed no man. But when, in language of the tenderest feeling, he spoke of his wife, whose highest passion had been gratified by the possession of a home so humble as even this – when he described how happy she had been in her garden, and how grief-stricken at being compelled to leave it – his eloquence fairly made my heart ache. I am sure my wife felt the full force of all he said. Her own attachment to the spot had already begun to take root, and she could sympathize with this rude sundering of a long-established tie.

CHAPTER V.
MAKING A PURCHASE –
FIRST IMPRESSIONS.

THE owner of these eleven acres had been for some months in the furnace of pecuniary affliction. He was going the way of nine-tenths of all the business flesh within the circle of my acquaintance. As a purchaser I did not seek him, nor to his representative did myself or my wife let fall a single word indicating that we were pleased with the property. When fifteen hundred dollars were named as the price, I did indulge in some expression of surprise, thinking it was quite enough. Discovering subsequently that the owner was an old city acquaintance, I dropped in one morning to see him, and for an hour we talked over the times, the markets, the savage rates demanded for money, and how the spring business was likely to turn out. On real estate I was mute as a mouse, except giving it as my decided opinion that some holders were asking greater prices than they would be likely to realize.

This side-thrust brought my friend out. He mentioned his house and eleven acres, and eagerly inquired if I did not know of some one who would buy. With as much indifference as I could assume, I asked his terms. He told me with great frankness that he was compelled to sell, and that his need of money was so great, that he might possibly do so whether the debtor got any thing or not. He urged me to find him a purchaser, and finally gave me the refusal of the place for a few days.

Now, the plain truth was, that my anxiety to buy was quite as great as his was to sell. During the next week we met several times, when he invariably inquired as to the prospect of a purchaser. But I had no encouragement to offer. When I thought I had fought shy long enough, I surprised him by saying that I knew of a purchaser who was ready to take the property at a thousand dollars. He sat down and indulged in some figuring, then for a few moments was silent, then inquired if the offer was a cash one, and when the money could be had. I replied, the moment his deed was ready for delivery.

It was evident that the offer of instant payment determined him to sell at so low a price – cash was every thing. Opening his desk, he took out a deed for the property, ready to execute whenever the grantee's name, the date and the consideration should have been inserted, handed it to me, and said he accepted the offer, only let him have the money as quickly as possible.

I confess to both exultation and surprise. I had secured an unmistakable bargain. The ready-made deed surprised me, but it showed the owner's necessities, and that he had been prepared to let the property go at the first decent offer. The natural selfishness of human nature has since induced me to believe that I could have bought for even less, had I not been so precipitate. His searches and brief of title were also ready: a single day or two was enough to bring them up – he had been determined to sell.

The transaction seemed to involve a succession of surprises. His turn for a new one came when he found that I had inserted my wife's name in the deed. So, paying him his thousand dollars, I returned with the deed to my wife, telling her that she had now a home of her own; that, come what might, the property was hers; that the laws of New Jersey secured it to her, and that no subsequent destitution of mine could wrest it from her. This little act of consideration was as gratifying a surprise to her as any that either buyer or seller had experienced. If rejoiced at my having secured the place, it gave to it a new interest in her estimation, and fixed and made permanent the attachment she had spontaneously acquired for it. Her gratification only served to increase my own.

It is thus that small acts of kindness make life pleasant and desirable. Every dark object is made light by them, and many scalding tears of sorrow are thus easily brushed away. When the heart is sad, and despondency sits at the entrance of the soul, a little kindness drives despair away, and makes the path cheerful and pleasant. Who then will refuse a kind act? It costs the giver nothing – just as this did; but it is invaluable to the receiver. No broader acres, no more stately mansion, whether in town or country, could now tempt my wife to leave this humble refuge. Here she has been ever happy, and here, I doubt not, she will end her earthly career.

In a week the house was vacated and cleansed, and we were in full possession. My wife was satisfied, my children were delighted, and I had realized the dream of twenty years ! One strong fact forced itself on my attention the first night I passed under my new roof. The drain of three hundred dollars per annum into the pocket of my city landlord had been stopped. My family received as safe a shelter for the interest of a thousand dollars, as he had given them for the interest of five thousand ! The feeling of relief from this unappeasable demand was indescribable. Curiously enough, my wife voluntarily suggested that the same feeling of relief had been presented to her. But in addition to this huge equivalent for the investment of a thousand dollars, there was that which might be hereafter realized from the cultivation of eleven acres of land.

This lodgment was effected on the first of April, 1855. When all our household fixings had been snugly arranged, and I took my first walk over my little plantation, on a soft and balmy morning, my feeling of contentment seemed to be perfect. I knew that I was not rich, but it was certain that I was not poor. In contrasting my condition with that of others, both higher and lower upon fortune's ladder, I found a thousand causes for congratulation, but none for regret. With all his wealth, Rothschild must be satisfied with the same sky that was spread over me. He cannot order a private sunrise, that he may enjoy it with a select circle of friends, nor add a single glory to the gorgeous spectacle of the setting sun. The millionaire could not have more than his share of the pure atmosphere that I was breathing, while the poorest of all men could have as much. God only can give all these, and to many of the poor he has thus given. All that is most valuable can be had for nothing. They come as presents from the hand of an indulgent Father, and neither air nor sky, nor beauty, genius, health, or strength, can be bought or sold. Whatever may be one's condition in life, the great art is to learn to be content and happy, indulging in no feverish longings for what we have not, but satisfied and thankful for what we have.

CHAPTER VI.
PLANTING A PEACH ORCHARD –
HOW TO PRESERVE PEACH-TREES.

It was now the season for me to bustle about, fix up my land, and get in my crops. I examined it more carefully, walked over it daily, and made myself thoroughly acquainted with it. As before mentioned, it had been utterly neglected for a whole season, and was grown up with enormous weeds. These, after a day or two of drizzling rain, when the seed-vessels were so wet as not to allow their contents to shatter out, I mowed off, gathered into several large heaps, and burned – thus getting rid of millions of pestiferous seeds. Then I purchased ploughs, including a subsoiler, a harrow, cultivator, and other tools. One acre of the whole was in clover, another was set aside as being occupied by the dwelling-house, garden, stable, and barn-yard; but much the larger half of that acre was allowed for garden purposes. This left me just nine acres for general fruit and vegetable culture. I hired a man to plough them up, he finding his own team, and another to follow him in the furrow with my subsoiler. The first went down ten inches, and the latter ten more.

My neighbors were extremely kind with their suggestions. They had never seen such deep ploughing, and warned me not to turn up the old subsoil, and thus bring it to the surface. But they were not book-farmers.

Now, this business of deep subsoil ploughing is a matter of indispensable value in all agriculture, but especially so in the planting of an orchard. No tree can thrive as it ought to, unless the earth is thoroughly and deeply loosened for the free expansion of the roots. If I could have ploughed two feet deep, it would have been all the better. In fact, the art of ploughing is in its mere infancy in this country. Too many of us follow blindly in the beaten track. The first plough was a tough, forked stick, of which one prong served as a beam, while the other dug the earth as a coulter. Of course the ploughing was only scratching. It would have been preposterous to expect the plough-man of Hesiod's or of Virgil's time to turn up and mellow the soil to a depth of fifteen or sixteen inches. Down to the present age, plough-ing was inevitably a shallow affair. But iron ploughs, steel ploughs, subsoil ploughs, have changed all this. It is as easy to-day to mellow the earth to the depth of two feet, as it was a century ago to turn over a sward to the depth of six inches. Besides, our fierce, trying climate, so different from the moist, milder one of England, Ireland, or even Holland, whence our ancestors emigrated, absolutely requires of us deep ploughing. Drought is our perpetual danger. Most crops are twenty to sixty per cent. short of what they would have been with adequate and seasonable moisture. That moisture exists not only in the skies above, but in the earth beneath our plants.

Though the skies may capriciously withhold it, the earth never will, if we provide a rich, mellow subsoil through which the roots can descend for moisture.

The hotter and dryer the weather, the better our plants will grow, if they have rich, warm earth beneath them, reaching down to and including moisture. We cannot, and we need not plough so very deep each year to assure this, if the subsoil is so underdrained that the su-perabundant moisture of the wet season does not pack it. Underd-raining as the foundation, and deep ploughing as the superstructure, with ample manuring and generous tillage, will secure us ample crops, such as any section of our country has rarely seen. Our corn should average seventy bushels per acre. Every field should be ready to grow wheat, if required. Every grass-lot should be good for three tons of hay per acre. Abundant fruits should gladden our fields and enrich our farmers' tables. So should our children no longer seek, in flight to crowded cities or the remote West, an escape from the ill-paid drudg-ery and intellectual barrenness of their fathers' lives, but find abun-dance and happiness in and around their childhood's happy homes.

I laid out two hundred dollars in the purchase OI old, well-rot-ted stable manure from the city, spread it over the ten acres, and ploughed up nine of them. I then set out my peach-trees on six acres, planting them in rows eighteen feet apart, and eighteen feet asunder in the rows. This accommodated a hundred and thirty-four to the acre, or eight hundred and four in all. These would not be in the way of any other crop, and in three years would be likely to yield a good return. The roots of every tree underwent a searching scrutiny before it was planted, to see that they harbored no members of that worm family which is so surely destructive of the peach. As trees are often delivered from the nursery with worms in them, so many of these were infected. The enemy was killed, and the butt of each tree was then swabbed with common tar, extending from where the roots begin to branch out, about twelve inches up. It is just about there, say between wind and water, at the surface of the ground, where the bark is soft, that in June and September the peach-moth deposits her eggs. From these is hatched the worm which kills the tree, unless picked out and destroyed.

To perform this searching operation on a thousand trees every year, would be laborious and expensive. There would also be great danger of its being imperfectly done, as many worms might escape the search, while the vital power of the tree would be seriously impaired by permitting them to prey upon its bark and juices even for a few months. Prevention would be far cheaper than curing. The offensive odor of the tar will cause the moth to shun the tree and to make her deposits somewhere else; while if any chance to light upon it, they will stick to the tar and there perish, like flies upon a sheet of flypaper.

The tar was occasionally examined during the season, to see that it kept soft and sticky; and where any hardening was discovered, a fresh swabbing was applied. The whole operation was really one of very little trouble, while the result was highly remunerative. Thoughtfulness, industry, and a little tar, did the business effectually. I believe no nostrum of putting ashes round the butt of a peach-tree to kill the worms, or any other nostrum of the kind, is worth a copper. The only sure remedy is prevention. Do not let the worms get in, and there will be no effort needed to get them out.

I planted none but the rarest and choicest kinds. Economy of a few cents in the price of a tree is no economy at all. It is the *best* fruit that sells the quickest and pays the highest profit. Yet there are still large quantities of fruit produced which is not worth taking to market. The best is cheaper for both buyer and seller. Hundreds of bushels of apples and peaches are annually made into execrable pies in all the large cities, merely because they can be purchased at less cost than those of a better quality. But it is a mistaken economy with the buyer, as a mild, good-flavored peach or apple requires less sugar, and will then make a better pie. Many persons have a pride in, and attach too much consequence to a tree which sprung up spontaneously on their own farm, or perhaps which they have cultivated with some care; and then numbers of comparatively worthless seedlings occupy the places that should be improved by finer varieties, and which, if cultivated, would afford a greater profit.

It is as easy to grow the choicest as the meanest fruit. I have a relative in Ohio who has a peach orchard of eleven acres, which has yielded him five thousand dollars in a single season, during which peaches were selling in Cincinnati at twenty-five cents a bushel. It is easy to understand that his orchard would not have produced him that sum at that price. No, it did not. He received two dollars a bushel more readily than his neighbors got twenty- five cents for the same variety of peaches, and this is how he did it. When the peaches had grown as large as a hickory nut, he employed a large force and put on one hundred and eighty-five days' work in picking off the excess of fruit. More than one- half of the fruit then upon the trees was carefully removed. Each limb was taken by hand, and where, within a space of eighteen inches, there would be probably twenty peaches, but six or seven of the fairest would be left to ripen. Thus, by carefully removing all but the strongest specimens, and throwing all the vigor of the tree into them, the peaches ripen early, and are remarkable for size and excellence of quality.

But this was labor ! Seven months' labor of one man in a small peach orchard ! But be it so – the net profit was between three and four thousand dollars. If he had neglected his trees, the owner's profits would have been a crop of peaches hardly fit to feed the pigs. I have profited largely by following his example, and will relate my own experience when the returns of my orchard come in.

I intend to be particular touching my peach orchard, as well for the gratification of my own pride, as an incentive to those who cannot be made to believe Ten Acres Enough. My success with it has far outstripped my expectations; and I pronounce a peach orchard of this size, planted and cultivated as it can be, and will be, by an intelligent man not .essentially lazy, as the sheet anchor of his safety. I was careful to plant none but small trees, because such can be removed from the nursery with greater safety than large ones, while the roots are less multiplied, and thus receive fewer injuries; neither are they liable to be displaced by high winds before acquiring a firm foothold in the ground. Many persons suppose that newly planted trees should be large enough to be out of danger from cattle running among them; but all cattle should be excluded' from a young orchard.

Moreover, small trees make a better growth, and are more easily trimmed into proper shape. All experienced horticulturists testify to the superior eligibility of small trees. They cost less at the nursery, less in transportation, and very few fail to grow. One year old from the bud is old enough, and the same, generally, may be said of apples and pears. I dug holes for each tree three feet square and two feet deep, and filled in with a mixture of the surrounding top-soil and leached ashes, a half bushel of the latter to each tree. Knowing that the peach- tree delights in ashes, I obtained four hundred bushels from a city soap-works, and am satisfied they were exactly the manure my orchard needed. Every root which had been wounded by the spade in removing the tree from the nursery, was cut off just back of the wound, paring it smooth with a sharp knife. The fine earth was settled around the roots by pouring in water; after which the mixture of earth and ashes was thrown on until the hole was filled, leaving a slight depression round the tree, to catch the rain, and the tree at about the same level it had maintained when standing in the nursery.

I did not stake up the trees. They were too small to need it; besides, I should be all the time on hand to keep them in position. Being a new-comer, I had no straw with which to mulch them, to retain the proper moisture about the roots, or it would have been applied. But the season turned out to be abundantly showery, and they went on growing from the start. Not a tree was upset by storm or wind, nor did one of them die. I do not think the oldest nurseryman in the country could have been more successful.

This operation made a heavy draft on the small cash capital which I possessed. But small as it was, it was large enough to show that capital is indispensable to successful farming. Had I been without it, my orchard would have been a mere hope, instead of a reality, and I might have been compelled to wait for years before feeling rich enough to establish it. But when the work of planting was over, my satisfaction was extreme; and when I saw the trees in full leaf, giving token that the work had been well done, I felt that I had not only learned but accomplished much. I had been constantly on the ground while the planting was progressing – had seen for myself that every tree was cleared of worms – had held them up while the water and the earth and ashes had been thrown in and gently packed about the roots – and had given so much attention in other ways, as to feel sure that no part of the whole operation had been neglected; and hence I had a clear right to regard it as my own job.

The cost of planting this orchard was as follows:

804 trees at 7 cents	$56.28
Planting them 2 cents	16.08
Ploughing and harrowing	20,00
400 bushels of ashes	48.00
Manure	200.00
	340.36

I have unfairly saddled on the orchard the whole charge of two hundred dollars for manure, because it went to nourish other crops which the same ground produced. But let that go – the land was quite poor, needed all it got, and I had no faith in farming without manure. Had my purse been heavy enough, the quantity should have been trebled.

As I am writing for the benefit of others, who, I hope, are not yet tired of peaches, let me add that this fruit will not succeed on ground where a previous orchard has been recently grown; neither can one be sure of getting healthy trees from any nurseryman who grows his on land from which he had recently produced a similar crop. The seed must be from healthy trees, and the buds from others equally free from disease. The peach, unless carefully watched and attended, is a short-lived tree. But it returns a generous income to a careful and generous grower. Of latter years the worm is its most formidable enemy. But with those who think a good tree is as much worth being taken care of as a good horse, there will be neither doubt nor difficulty in keeping the destroyer out.

Ten well-grown, bearing trees, which I found in the garden, were harboring a hundred and ninety worms among them when I undertook the work OI extermination. I bared the collar and roots of each tree as far as I could track a worm, and cut him out. I then scrubbed the whole exposed part with soapsuds and a regular scrubbing-brush; after which I let them remain exposed for a week. If any worms had been overlooked, the chips thrown out by their operations would be plainly visible on the clean surface at the week's end. Having tracked and cut out them also, I felt sure the enemy was exterminated, and covered up the roots, but first using the swab of common tar, applying it all round the collar, and some distance up.

These garden-trees were terribly scarified by the worms. But the cleaning out I gave them was effectual. The soap-suds purged the injured parts of the unhealthy virus deposited by the worms, leaving them so nice and clean that the new bark began immediately to close over the cavities, and soon covered them entirely. I thus saved ten valuable bearing trees. Then I shortened in the long, straggling branches, for the peach will certainly grow sprawling out on every side, forming long branches which break down under the weight of a full crop at their extremities, unless the pruning-knife is freely used every season. All this was the work of less than a day, and shows that if peach-orchards perish after bearing only two or three crops, it may be attributed solely to mere neglect and laziness on the part of their owners. They plant trees, refuse to take care of them, and then complain if they die early. The world would soon be without pork, if all the pigs were as much neglected. These ten trees have never failed to produce me generous crops of luscious fruit. I cannot think of any investment which has paid me better than the slight labor annually required to keep them in good condition.

I have tried with entire success two other methods of protecting peach-trees from the ravages of the worm. I have found gas-tar equally effectual with the common tar, and much more easily obtained. But care must be taken not to cover a height of more than four to six inches of the butt of the tree. If the whole stem from root to branch be covered, the tree will surely die. Another method is to inclose the butt in a jacket of pasteboard, or even thick hardware paper, keeping it in place with a string, and lowering it an inch or two below the ground, so as to prevent the fly having access to the soft part of the bark. These jackets will last two or three years, as they should be taken off at the approach of winter, to prevent them from becoming a harbor for insects. But they are an infallible preventive. I have recently procured a supply of the thick tarred felt which is used for making paper roofs, to be cut up and turned into jackets. This material will last for years, being water-proof, while the odor of the gastar in which it has been steeped is peculiarly offensive to the whole tribe of insects.

CHAPTER VII.
PLANTING RASPBERRIES AND STRAWBERRIES – TRICKS OF THE NURSERY.

My peach-orchard was no sooner finished than I filled each row with raspberries, setting the roots two feet apart in the rows. This enabled me to get seven roots in between every two trees, or five thousand six hundred and fifty-six in all. This was equivalent to nearly two acres wholly planted with raspberries according to the usual plan. They would go on growing without injuring the peach-trees, or being injured by them; and when the latter should reach their full growth, their shade would be highly beneficial to the raspberries, as they thrive better and bear more freely when half protected from the burning sun. The tops were cut off within a few inches of the ground, thus preventing any excessive draft upon the newly planted roots. No staking up was needed. These roots cost me six dollars per thousand, or thirty-four dollars for the lot, and were the ordinary Red Antwerp. The season proving showery, they grew finely. Some few died, but my general luck was very satisfactory. I planted the whole lot in three days with my own hands.

I am sure the growth of my raspberries was owing, in a great degree, to the deep ploughing the land had received. The soil they delight in is one combining richness, depth, and moisture. It is only from such that a full crop may he expected every season. The roots must have abundance of elbow-room to run down and suck up moisture from the abundant reservoir which exists below. Deep ploughing will save them from the effects of dry weather, which otherwise will blast the grower's hopes, giving him a small berry, shrivelled up from want of moisture, instead of one of ample size, rich, and juicy. Hence irrigation has been known to double the size of raspberries, as well as doubling the growth of the canes in a single season. Mulching also is a capital thing. One row so treated, by way of experiment, showed a marked improvement over all the others, besides keeping down the weeds.

As a market fruit the raspberry stands on the same list with the best, and I am satisfied that one cannot produce too much. For this purpose I consider the Red Antwerp most admirably adapted. There are twenty other varieties, some of which are probably quite as valuable, but I was unwilling to have my attention divided among many sorts. One really good berry was enough for me. Some of my neighbors have as much as ten acres in this fruit, from which they realize prodigious profits. Like all the smaller fruits, it yields a quick return to an industrious and pains-taking cultivator.

Immediately on getting my raspberries in, I went twice over the six acres with the cultivator, stirring up the ground some four inches deep, as it had been a good deal trampled down by our planting operations. This I did myself, with a thirty-dollar horse which I had recently bought. Having eighteen feet between two rows of peach-trees, I divided this space into five rows for strawberries, giving me very nearly three feet between each row. In these rows I set the strawberry plants, one foot apart, making about 10,000 plants per acre, allowing for the headlands. I bought the whole 60,000 required for §2 per thousand, making §120. This was below the market price.

In planting these I got three of the children to help me, and though it was more tiresome work than they had ever been accustomed to, yet they stood bravely up to it. Every noon we four went home with raging appetites for dinner, where the plain but well-cooked fare provided by my wife and eldest daughter – for she kept no servant – was devoured with genuine country relish. The exercise in the open air for the whole week which it took us to get through this job did us all a vast amount of good. Roses came into the cheeks of my daughters, to which the cheeks aforesaid had been strangers in the city; and it was the general remark among ns at breakfast, that it had never felt so good to get to bed the night before. Thus honest labor brought wholesome appetites and sound repose. Most of us complained of joints a little stiffened by so much stooping, but an hour's exercise at more stooping made us limber for the remainder of the day.

It occupied us a whole week to set out these plants, for we were all new hands at the business. But the work was carefully done, and a shower coming on just as we had finished, it settled the earth nicely to the roots, and I do not think more than two hundred of them died. I intended to put a pinch of guano compost or a handful of poudrette into each hill, but thought I could not afford it, and so let them go, trusting to being able to give them a dressing of some kind of manure the following spring. I much regretted this omission, as I was fully aware of the great value of the best strawberries, and plenty of them. My wife thought at first that six acres was an enormous quantity to have – inquired if I expected to feed the family on strawberries, and whether it was not worth while to set about raising some sugar to go with them, feeling certain that a great deal of *that* would be wanted.

I forgot to say that I had planted Wilson's Albany Seedling. This was the berry for which we had been compelled to pay such high prices while living in the city. Everybody testified to its being the most profuse bearer, while its great size and handsome shape made it eagerly sought after in the market. It was admitted, all things con-sidered, to be the best market berry then known. My experience has confirmed this. True, it is a little tarter than most other varieties, and therefore requires more sugar to make it palatable; but this objection is more theoretical than practical, as I always noticed that when the berries came upon the table, while living in the city, we continued to pile on the sugar, no matter what the price or quantity. The berries were there, and must be eaten.

On one occasion, on repeating this observation to my wife, she admitted having noticed the same remarkable fact, and added that she believed strawber ries would continue to be eaten, even if each quart required a pound of sugar to sweeten it. She declared that for her part, she and the children intended to do so in future.

Now, although she was extravagantly fond of strawberries, and had brought up our children in the same faith, this threat did not alarm me, for I knew that hereafter our berries would cost me nothing, and that if they devoured them too freely, sugar included, a slight pain under the apron of some of them would be likely to moderate their infatuation. I then suggested to her, how would it do – whether it would not make our establishment immensely popular – if in selling my berries, when the crop came in next year, to announce to the public that we would throw the sugar in? She looked at me a moment, and must have suspected that I was quizzing her; for she got up and left the room, saying she must go into the kitchen, as she heard the tea-kettle boiling over. But though I waited a full half hour, yet she did not return.

The reader may have been all this time watching the condition of my purse. But he has not been so observant as myself. These plants did not cost me cash. I had intended to plant an acre or two to begin with. But after buying my peach-trees and raspberries, the nurseryman inquired if I did not intend to plant strawberries also, as he had a very large quantity which he would sell cheap. His saying that he had a very large lot, and that he would sell them cheap, seemed to imply that he found a difficulty in disposing of them. Besides, the selling season was pretty nearly over. I therefore fought shy, and merely inquired his terms. This led to a long colloquy between us, in the course of which I held off just in proportion as he became urgent. At last, believing that I was not disposed to buy, although I went there for that very purpose, he offered to sell me 60,000 plants for $120, and to take his money out of the proceeds of my first crop. This offer I considered fair enough, much better than I expected; and after having distinctly agreed that he should depend upon the crop, and not on me, for payment, and that if the coming season yielded nothing he should wait for the following one, I confessed to him that his persuasions had overcome me, and consented to the bargain.

In other words, I did not run in debt – I saved just that much of my capital, and could make a magnificent beginning with our favorite fruit. As I was leaving this liberal man, he observed to me:

"Well, I am glad you have taken this lot, as I was intending to plough them in to-morrow."

"How is that?" I inquired, not exactly understanding his meaning.

"Oh," said he, "I have so many now that I must have the ground for other purposes, and so meant to plough them under if you had not bought them."

This was an entirely new wrinkle to me, and fully explained why he could afford to farm them out on the conditions referred to. Though a capital bargain for me, yet it was a still better one for him. What he was to receive was absolutely so much clear gain.

But then, after all that has been said and written, is it not a truth that cannot be disputed, that no bargain can be pronounced a good one unless all the parties to it are in some way benefited?

Here, now, were six acres of ground pretty well crowded up, at least on paper. But the strawberries would never grow higher than six inches; the raspberries would be kept down to three or four feet, while the peaches would overtop all. Each would be certain to keep out of the other's way. Then look at the succession. The strawberries would be in market first, the raspberries would follow, and then the peaches, for of the latter I had planted the earliest sorts; so that, unlike a farm devoted wholly to the raising of grain, which comes into market only once a year, I should have one cash-producing crop succeeding to another during most of the summer. On the remaining three acres I meant to raise something which would bring money in the autumn, so as to keep me flush all the time. You may say that this was reckoning my chickens before they were hatched; but you will please remember that thus far I have not even mentioned chickens, and I pray that you will be equally considerate. I know, at least I have some indistinct recollection of having heard that the proof of the pudding lay in the eating. But pray be patient, even credulous, until the aforesaid mythical pudding is served up. I am now cooking it, and you ought all to know that cooks must not be hurried. In good time it will come smoking on the table.

CHAPTER VIII.
BLACKBERRIES –
A REMARKABLE COINCIDENCE.

In the course of my agricultural reading for some years previous to coming into the country, I had noticed great things said of a new blackberry which had been discovered in the State of New York. The stories printed in relation to it were almost fabulous. It was represented as growing twenty feet high, and as bearing berries nearly as large as a walnut, which melted on the tongue with a lusciousness to which the softest ice-cream was a mere circumstance, while the fruit was said to be strung upon its branches like onions on a rope. A single bush would supply a. large family with fruit ! I was amazed at the extravagant accounts given of its unexampled productiveness and matchless flavor. I had supposed that I knew all about blackberries, but here was a great marvel in a department which had been proverbially free from eccentricities of that kind.

But I followed it – in the papers – for a long time. At last I saw it stated that the rare plant could not be propagated from the seed, but only from suckers, and therefore very slowly. Of course it could not be afforded for less than a dollar apiece ! It would be unreasonable to look for blackberries for less ! It struck me that the superior flavor claimed for it must be a little of the silvery order – that in berries bought at that price, a touch might be detected even of the most auriferous fragrance. Still, I was an amateur – in a small way. I rejoiced in a city garden 'which would readily accommodate a hundred of this extraordinary berry, especially as it was said to do better and bear more fruit, when cut down to four feet, instead of being allowed to grow to a height of twenty.

It thus seemed to be made for such miniature gardeners as myself. One generous advertiser offered to send six roots by mail for five dollars, provided ten red stamps were inclosed with the money. I had never before heard of blackberries being sent by mail; but the whole thing was recommended by men in whose standing all confidence could be placed, and who, as far as could be discovered, had no plants to sell. Under such circumstances, doubt seemed to be absurd.

I sent five dollars and the stamps. But this was one of the secrets I never told my wife until she had eaten the first bowlful of the fully ripened fruit, eighteen months afterwards. Well, the plants came in a letter – mere fibres of a greater root – certainly not thicker than a thin quill, not one of them having a top. They looked like long white worms, with here and there a bud or eye. I never saw, until then, what I considered the meanest five dollars' worth of any thing I had ever bought; and when my wife inquired what those things were I was planting, I replied that they were little vegetable wonders which a distant correspondent had sent me. Not dreaming that they cost me near a dollar apiece, at the very time I owed a quarter's rent, she dropped the subject.

But I planted them in a deeply spaded and rich sunny border, deluged them every week with suds from the family wash, and by the close of the season they had sent up more than a dozen strong canes which stood six feet high. The next summer they bore a crop of fruit which astonished me. From the group of bushes I picked fifteen quarts of berries superior to any thing of the kind we had ever eaten. I then confided the secret to my wife: she considered the plants cheap at five dollars, and pronounced my venture a good one. I think we had more than five dollars' worth of satisfaction in show-ing them to our friends and neighbors. We gave away some pints of the fruit, and such was its fame and popularity, that I feel convinced we could have readily disposed of it all in the same way.

One of the reporters for a penny-paper hearing of the matter, called in my absence to see them. My wife politely acted as showman, and being very eloquent of speech on any matter which happens to strike her fancy, she was quite as communicative as he desired. She did not know that the fellow was a penny-a-liner, whose vocation it was to magnify an ant-hill into a mountain. To her extreme consternation, as well as to mine, the next morning's paper contained a half-column article describing my blackberries, even giving my name and the number of the house. By ten o'clock that day the latter was run down with strangers, who had thus been publicly invited to call and see the new blackberry. Our opposite neighbors laughed heartily over my wife's vexation, and for the first time in my life I saw her almost immovable good temper give way. The nuisance continued for weeks, as the vile article had been copied into some of the neighboring country papers, and thus new swarms of bores were inflamed with curiosity. This little vexatious circumstance afforded unmistakable evidence of the great interest taken by the public in the discovery of a new and valuable fruit. I could have disposed of thousands of plants if I had had them for sale.

This was the New Rochelle or Lawton Blackberry. The numerous suckers which came up around each root I transplanted along my border, until I had more than two hundred of them. This was long before a single berry had been offered for sale in the Philadelphia market, though the papers told me that the fruit was selling in New York at half a dollar per quart, and that the great consuming public of that city, having once tasted of it, was clamorous for more. I am constrained to say that the nurserymen who had these plants to sell did not over praise them. This berry has fully realized all they promised in relation to it; and a debt of thankfulness is owing to the men who first discovered and caused it to be propagated. It has taken its place in public estimation beside the strawberry and raspberry, and will henceforth continue to be a favorite in every market where it may become known.

This extraordinary fruit was first noticed in 1834, by Mr. Lewis A. Secor, of New Rochelle, New York, who observed a single bush growing wild in an open field, but loaded with astonishing clusters of larger berries than he had ever seen, and of superior richness of flavor. At the proper season he removed the plant to his garden, where he continued to propagate it for several years, during which time it won the unqualified admiration of all who had an opportunity of either seeing or tasting the fruit. Numerous plants were distributed, and its propagation in private gardens and nurseries began. A quantity of the fruit being exhibited at the Farmers' Club, by Mr. William Lawton, the club named it after him, leaving the discoverer unrecognized.

Great sums of money have been made by propagators of this berry. It possesses peculiar merits in the estimation of market gardeners. It ripens just as the supply of strawberries and raspberries has been exhausted, and before peaches and grapes have made their appearance, filling with delicious fruit a horticultural vacuum which had long existed. Its mammoth size and luscious qualities insure for it the highest prices, and it has steadily maintained its original character. It pays the grower enormously, is a sure bearer, is never touched by frost or attacked by insect enemies, and when well manured and staked up from the wind, and cut down to four feet high, with the limbs shortened to a foot, will readily produce two thousand quarts to the acre. Some growers have greatly exceeded this quantity. I have known a single plant to yield eighteen hundred berries, and three plants to produce sixteen quarts. Its flavor is entirely different from that of the common wild blackberry, while it .abounds in juice, and contains no core. It is evidently a distinct variety. It has also long been famous for yielding a most superior wine.

When I went into the country I had two hundred of the Lawton blackberry to plant, all which were the product of my five-dollar venture. In digging them up from my city garden, every inch of root that could be found was carefully hunted out. They had multiplied under ground to a surprising extent – some of them being as much as twenty feet in length. These roots were full of buds from which new canes would spring. Their vitality is almost unconquerable – everybody knows a blackberry is the hardest thing in the world to kill. I cut off the canes six inches above the root, then divided each stool into separate roots, and then cutting up the long roots into slips containing one to two eyes each, I found my number of sets to exceed a thousand, quite enough to plant an acre.

These I put out in rows eight feet apart, and eight feet asunder in the rows. Not ten of them died, as they came fresh out of the ground in one place, only to be immediately covered up some three inches deep in another. Thus this whole five-dollar speculation was one of the luckiest hits I ever made; because I began early, before the plant had passed into everybody's hands; and when it came into general demand, I was the only grower near the city who had more than a dozen plants. Very soon everybody wanted the fruit, and the whole neighborhood wanted the plants. How I condescended to supply both classes of customers will appear hereafter.

Yet, while setting out these roots, several of my neighbors, as usual when I was doing any thing, came to oversee me. On former occasions they had expressed considerable incredulity as to my operations; and it was easy to see from their remarks and inquiries now, that they thought I didn't know much, and would have nothing for my labor but my pains. I always listened good-humoredly to their remarks, because I discovered that now and then they let fall something which was of real value to me. When they discovered it was blackberries I was planting, some of them laughed outright. But I replied that this Lawton berry was a new variety, superior to any thing known, and an incredible bearer. They answered me they could find better ones in any fence corner in the township, and that if I once got them into my ground I could never get them out. It struck me the last remark would also apply as justly to my peach-trees.

But I contented myself with saying that I should never want to get them out, and that the time would come when they would all want the same thing in their own ground. Thus it is that pioneers in any thing are generally ridiculed and discouraged by the general multitude. Of all my visitors, only two appeared to have any correct knowledge of the new plant. They offered to buy part of my stock; but on refusing to sell, they engaged to take some in the autumn.

I have been thus particular in writing of the Lawton, because of my singular success with it from the start. I thus occupied my seventh acre; but the rows being eight feet apart, abundant room was left to raise a crop of some kind between them. Even in the rows, between the roots, I planted corn, which grew well, and afforded a most beneficial shade to the young blackberries as they grew up. I am satisfied they flourished better for being thus protected the first season from the hot sun. When in full maturity, they need all the sun they can get. They will grow and flourish in almost any soil in which they once become well rooted, though they are rank feeders on manure. Like a young pig, feed them well and they will grow to an astonishing size: starve them, and your crops will be mere runts. It is from the same skinning practice that so many corn-cribs are seen to abound in nubbins.

I had thus two acres left unoccupied; one acre, as previously stated, was most fortunately in clover. On this I put four bushels of ground plaster mixed with a sprinkling of guano, the two costing me only five dollars. I afterwards devoted an acre to tomatoes, and the last to parsnips, cabbages,* turnips, and sweet corn. This latter was scattered in rows or drills three feet apart, intending it for green fodder for the horse and cow when the clover gave out. The turnips were sowed between the com-rows, and were intended for winter feeding for horse and cow. On the acre of blackberries, between the rows, I planted cabbage, putting into each hill a spoonful of mixed plaster and guano, and wherever I could find vacant spots about the place, there also a cabbage plant was set out. A few pumpkin hills were started in suitable places. In fact, my effort was to occupy every inch of ground with something. The cabbage and tomato plants cost me thirty dollars.

These several crops were put in as the season for each one came round. The green-corn crop was not all put in at one time, but at intervals about two weeks apart, so that I should have a succession of succulent food during the summer. The horse and cow were to be kept in the barnyard, as I had no faith in turning cattle out to pasture, thus requiring three times as much land as was necessary, besides losing half the manure. The latter was a sort of hobby with me. I was determined to give my crops all they could profitably appropriate, and so soil my little stock; that is, keep them in the barnyard in summer, and in the stable in winter, while their food was to be brought to them, instead of their being forced to go after it. I knew it would cost time and trouble; but I have long since discovered that most things ot value in this world come to us only as the result of diligent, unremitted labor. The man, even upon ten acres, who is content to see around him only barren fields, scanty crops, and lean, starving animals, does not deserve the name of farmer. Unless he can devise ways and means for changing such a condition of things, and cease ridiculing all propositions of amendment that may be pointed out to him, he had better be up and off, and give place to a live man. Such skinning and exhausting tillage is one cause or the annual relative decline of the wheat-crop all over the Union, and of the frequent changes in the ownership of lands. The fragrance of a fat and ample manure heap is as grateful to the nostrils of a good farmer, as the fumes of the tavern are notoriously attractive to those of a poor one.

CHAPTER IX.
THE GARDEN – FEMALE MANAGEMENT – COMFORTS AND PROFITS.

I MENTIONED some time ago that the wife of the former owner of this place had left it with a world of regrets. She had been passionately fond of the garden which now fell to us. As daylight can be seen through very small holes, so little things will illustrate a person's character. Indeed, character consists in little acts, and honorably performed; daily life being the quarry from which we build it up and rough-hew the habits that form it. The garden she had prepared, and cultivated for several years, doing much of the work of planting, watching, watering, and training with her own hands, bore honorable testimony to the goodness of hers. She had filled it with the choicest fruit-trees, most of which were now in full bearing. There was abundance of all the usual garden fruits, currants, gooseberries, grapes, and an ample asparagus bed. It was laid out with taste, convenience, and liberality. Flowers, of course, had not been omitted by such a woman. Her vocation had evidently been something beyond that of merely cooking her husband's dinners. But her garden bore marks of long abandonment. Great weeds were rioting in the borders, grass had taken foothold in the alleys, and it stood in need of a new mistress to work up into profitable use the store of riches it contained. It struck me that if one woman could establish a garden like this, I could find another on my own premises to manage it.

After I had got through with the various plantings of my standard fruits – indeed, while much of it was going on – I took resolute hold of the garden. It was large enough to provide vegetables for three families. I meant to make it sure for one. With all the lights and improvements of modern times, and they are many, three-fourths of the farm gardens in our country are still a disgrace to our husbandry. As a rule, the most easily raised vegetables are not to be found in them; and the small fruits, with the exception of currants and goose-berries, are universally neglected. Many of our farmers have never tasted an early York cabbage. If they get cabbages or potatoes by August, they think they are doing pretty well. They do not understand the simple mysteries of a hot-bed, and so force nothing. Now, with this article, which need not cost five dollars, and which a boy of ten years can manage, you can have cabbages and potatoes in June, and beans, tomatoes, cucumbers, and squashes, and a host of other delicious vegetables, a little later.

By selecting your seed, you can have salad, green peas, onions, and beets by the last of June, or before, without any forcing. A good asparagus bed, covering two square rods of ground, is a luxury that no farmer should be without. It will give him a palatable dish, green and succulent from the bosom of the earth every day, from May to July. A good variety of vegetables is within the reach of every fanner the year round. They are not only an important means of supporting the family, paying at least one-half the table expenses, but they are greatly conducive to health. They relieve the terrible monotony of salt junk, and in the warm season prevent the fevers and bowel complaints so often induced by too much animal food.

Neglect is thus too much the rule. A row of currants, for example, is planted in a garden. It will indeed bear well with neglect; but an annual manuring and thinning out of old wood, would at least triple the size of the fruit,- and improve its quality. The row of currants will furnish a daily supply of refreshing fruit to the table for months together. Why should its culture then be totally neglected, when a row of corn by its side of equal length, which will supply only a single feeding to a pen of hogs, is most carefully manured, watched, ploughed, and hoed? I have sometimes seen fanners who, after expending large sums in establishing a young orchard of trees, would destroy one-half by choking them with a crop of oats or clover, because they could not afford to lose the use of the small strip of land a few feet wide in the row, which ought to have been kept clean and cultivated.

I began by deepening the garden soil wherever a spade could be put in. I hired a man for this purpose, and paid him ten dollars for the job, including the hauling and digging in of the great pile of manure I had found in the barnyard, and the clearing up of things generally. I would have laid out fifty dollars in manure, if the money could have been spared; but what I did afforded an excellent return. My wife and eldest daughter, Kate, then in her eighteenth year, did all the planting. I spent five dollars in buying for them a complete outfit of hoes, rakes, and trowels for garden use, lightly made on purpose for female handling, with a neat little wheelbarrow to hold the weeds and litter which I felt pretty sure would have to be hoed up and trundled away before the season was over.

They took to the garden manfully. I kept their hoes constantly sharpened with a file, and they declared it was only pastime to wage warfare on the weeds with weapons so keen. Now and then one of the boys went in to give them a lift; and when a new vegetable bed was to be planted, it was dug up and made ready for them. But the great bulk of all other work was done by themselves.

Never has either of them enjoyed health so robust, or appetites so wholesome. As a whole year's crop of weeds had gone to seed, they had millions of the enemy to contend with, just as I had anticipated. I did not volunteer discouragements by repeating to them the old English formula, that

> "One year's seeding
> Makes seven years' weeding,"

but commended their industry, exhorted them to persevere, and was lavish in my admiration of the handsome style in which they kept the grounds. I infused into their minds a perfect hatred of the whole tribe of weeds, enjoined it upon them not to let a single one escape and go to seed, and promised them that if they thus exterminated all, the next year's weeding would be mere recreation.

I will say for them, that all our visitors from the city were surprised at seeing the garden so free from weeds, while they did not fail to notice that most of the vegetables were extremely thrifty. They did not know that in gardens where the weeds thrive undisturbed, the vegetables never do. As to the neighbors, they came in occasionally to see what the women were doing, but shook their heads when they saw they were merely hoeing up weeds – said that weeds did no harm, and they might as well attempt to kill all the flies – they had been brought up among weeds, knew all about them, and " it was no use trying to get rid of them."

But the work of weeding kept on through the whole season, and as a consequence, the ground about the vegetables was kept constantly stirred. The result of this thorough culture was, that nearly every thing seemed to feel it, and the growth was prodigious, far exceeding what the family could consume. We had every thing we needed, and in far greater abundance than we ever had in the city. I am satisfied this profusion of vegetables lessened the consumption of meat in the family one-half. Indeed, it was such, that my wife suggested that the garden had so much more in it than we required, that perhaps it would be as well to send the surplus to the store where we usually bought our groceries, to be there sold for our benefit.

The town within half a mile of us contained some five thousand inhabitants, among whom there was a daily demand for vegetables. I took my wife's advice, and from time to time gathered such as she directed, for she and Kate were sole mistresses of the garden, and sent them to the store. They kept a regular book-account of .these consignments, and when we came to settle up with the storekeeper at the year's end, were surprised to find that he had eighty dollars to our credit. But this was not all from vegetables – a good deal of it came from the fruit-trees.

After using in the family great quantities of fine peaches from the ten garden-trees, certainly three times as many as we could ever afford to buy when in the city, the rest went to the store. The trees had been so hackled by the worms that they did not bear full crops, yet the yield was considerable. Then there were quantities of spare currants, gooseberries, and several bushels of common blue plums, which the curculio does not sting. When my wife discovered there was so ready a market at our own door, she suffered nothing to go to waste. It was a new feature in her experience – every thing seemed to sell. Whenever she needed a new dress for herself or any of the children, all she had to do was to go to the store, get it, and have it charged against her garden fund. I confess that her success greatly exceeded my expectations.

Let me now put in a word as to the cause of this success with our garden. It was not owing to our knowledge of gardening, for we made many blunders not here recorded, and lost crops of two or three different things in consequence. Neither was it owing to excessive richness of the ground. But I lay it to the unsparing warfare kept up upon the weeds, which thus prevented their running away with the nourishment intended for the plants, and kept the ground constantly stirred up and thoroughly pulverized. I have sometimes thought one good stirring up, whether with the hoe, the rake, or the cultivator, was as beneficial as a good shower.

When vegetables begin to look parched and the ground becomes dry, some gardeners think they must commence the use of the watering-pot. This practice, to a certain extent, and under some circumstances, may perhaps be proper, but as a general rule it is incorrect. The same time spent in hoeing, frequently stirring the earth about vegetables, is far preferable. When watering has once commenced it must be continued, must be followed up, else you have done mischief instead of good; as, after watering a few times, and then omitting it, the ground will bake harder than if nothing had been done to it. Not so with hoeing or raking. The more you stir the ground about vegetables, the better they are off; and whenever you stop hoeing, no damage is done, as in watering. Vegetables will improve more rapidly, be more healthy, and in better condition at maturity, by frequent hoeing than by frequent watering. This result is very easily shown by experiment. Just notice, after a dewy night, the difference between ground lately and often stirred, and that which has lain unmoved for a long time. Or take two cabbage plants under similar circumstances; water one and stir the other just as often, stirring the earth about it carefully and thoroughly, and see which will distance the other in growth.

There are secrets about this stirring of the earth which chemists and horticulturists would do well to study with the utmost scrutiny and care. Soil cultivated in the spring, and then neglected, soon settles together. The surface becomes hard, the particles cohere, they attract little or no moisture, and from such a surface even the rain slides off, apparently doing little good. But let this surface be thoroughly pulverized, though it be done merely with an iron rake, and only a few inches in depth, and a new life is infused into it. The surface becomes friable and soft, the moisture of the particles again becomes active, attracting and being attracted, each seeming to be crying to his neighbor, "Hand over, hand over – more drink, more drink." Why this elaboration should grow less and less, till in a comparatively short time it should seem almost to cease, is a question of very difficult solution; though the varying compositions of soils has doubtless something to do with the matter.

But let the stirring be carefully repeated, and all is life again. Particles attract moisture from the atmosphere, hand it to each other, down it goes to the roots of vegetables, the little suction fibres drink it in; and though we cannot see these busy operations, yet we perceive their healthy effects in the pushing up of vegetables above the surface. The hoe is better than the water-pot. My garden is a signal illustration of the fact.

CHAPTER X.
CHEATED IN A COW – A GOOD AND A BAD ONE – THE SAINT OF THE BARNYARD.

BOTH myself and wife had always coveted a cow. All of the family were extravagantly fond of milk. Where so many children were about, it seemed indispensable to have one; besides, were we not upon a farm? and what would a farm be without having upon it at least one saint of the barnyard? As soon as we came on the place, I made inquiries of two or three persons for a cow. The news flew round the neighborhood with amazing rapidity, and in the course of two weeks I was besieged with offers. They haunted me in the street, as I went daily to the post-office; even in the evening, as we sat in our parlor. It seemed as if everybody in the township had a cow to sell. Indeed, the annoyance continued long after we had been supplied.

Now, though I knew a great deal of milk, having learned to like it the very day I was born, yet I was utterly ignorant of how to choose a cow, and at that time had no friend to advise with. But I suspected that no one who had a first-rate animal would voluntarily part with it, and so expected to be cheated. I hinted as much to my wife, whereupon she begged that the choice might be left to her; to which I partially consented, thinking that if we should be imposed on, I should feel better if the imposition could be made chargeable somewhere else than to my own ignorance. Besides, I knew that she could hardly be worse cheated than myself.

One morning a very respectable-looking old man drove a cow up to the door, and called us out to look at her. My wife was pleased with her looks the moment she set eyes on her, while the children were delighted with the calf, some two weeks old. I did not like her movements – she seemed restless and ill-tempered; but the old man said that was always the way with cows at their first calving. Still, I should not have bought her. But somehow my wife seemed bewitched in her

JAMES MILLER

favor, and was determined to have her. This the old man could not fail to notice, and was loud in extolling her good qualities, declaring that she would give twenty quarts of milk a day. After some further parley, he inadvertently admitted that she had never been milked. My wife did not notice this striking discrepancy of a cow giving twenty quarts daily, when as yet no one had ever milked her; but the lie was too bouncing a one to escape my notice. As I saw my wife had set her heart upon the cow, I said nothing, and finally bought cow and calf for thirty dollars, though quite certain they could have been had for five dollars less, if my wife had not so plainly shown to the old sinner that she was determined to have them. I do not think she will ever be up to me in making a bargain. But as it had been agreed that she should choose a cow, so she was permitted to have her own way.

At the end of the week the calf was sold for three dollars – a low price; but then my wife wanted the milk, and she and Kate were anxious to begin the milking. I am sure I was quite willing they should have all they could get. When they did begin, there was a great time. Now, most women profess to understand precisely how a cow should be milked, and yet comparatively few know any thing about it. They remind me of the Irish girls who are hunting places. These are all first-rate cooks, if you take their word for it, and yet not one in a hundred knows any thing of even the first principles of cooking.

The first process in the operation of milking is to fondle with the cow, make her acquaintance, and thus give her to understand that the man or maid with the milking pail approaches her with friendly intentions, in order to relieve her of the usual lacteal secretion. It will never do to approach the animal with combative feelings and intentions. Should the milker be too impetuous; should he swear, speak loud and sharp, scold or kick, or otherwise abuse or frighten the cow, she will probably prove refractory as a mule, and may give the uncouth and unfeeling milker the benefit of her heels, – a very pertinent reward, to which he, the uncouth milker, is justly entitled. Especially in the case of a new milker, who may be a perfect stranger to the cow, the utmost kindness and deliberation arc necessary.

Before commencing to milk, a cow should be fed, or have some kind of fodder offered her, in view of diverting her attention from the operation of milking. By this means the milk is not held up, as the saying is, but is yielded freely. All these precautions are more indispensable when the cow has just been deprived of her calf. She is then uneasy, fretful, and irritable, and generally so disconsolate as to need the kindest treatment and the utmost soothing. The milker should be in close contact with the cow's body, for in this position, if she attempt to kick him, he gets nothing more than a push, whereas if he sits off at a distance, the cow has an opportunity to inflict a severe blow whenever she feels disposed to do so.

All milkers of cows should understand that the udder and teats are highly organized, and consequently very sensitive; and these facts should be taken into consideration by amateur milkers, especially when their first essay is made on a young animal after the advent of her first calf, and that one just taken from her. At this period, the hard tugging and squeezing to which many poor dumb brutes have to submit in consequence of the application of hard-fisted, callous, or inexperienced fingers, is a barbarity of the very worst kind; for it often converts a docile creature into a vicious one, from which condition it is extremely difficult, if not impossible, to wean her.

Of every one of these requisites both wife and daughter were utterly ignorant. They went talking and laughing into the barn, one with a bright tin pail in her hand, an object which the cow had never before seen, and both made at her, forgetting that they were utter strangers to her. Besides, she was thinking of her absent calf, and did not want to see any thing else. Their appearance and clamor of course frightened her, and as they approached her, so she avoided them. They followed, but she continued to avoid, and once or twice put down her head, shook it menacingly, and even made an incipient lunge at them with her sharply pointed horns. These decided demonstrations of anger frightened them in turn, and they forthwith gave up the pursuit of milk in the face of difficulties so unexpected. We got none that night. In the morning we sent for an experienced milker, but she had the utmost difficulty in getting the cow to stand quiet even for a moment. My wife was quite subdued about the matter. It would never do to keep a cow that nobody could milk She said but little, however – it was *her* cow. Longer trial produced no more encouraging result, as she seemed untamable, and my wife was glad to have me sell her for twenty dollars, at the same time resolving never again to buy a cow with her first calf.

It was voted unanimously that another should be procured, and that this time the choice should be left to me. Now, I never had any idea of buying poor things of any kind merely because they were cheap. When purchasing or making tools or machinery, I never bought or made any but the very best, as I found that even a good workman could never do a good job with poor tools. So with all my farm implements – I bought the best of their kind that could be had. If my female gardeners had been furnished with heavy and clumsy hoes and rakes, because such were cheap, their mere weight would have disgusted them with the business of hoeing and weeding. So with a cow. It is true, I had become the owner of a magnificent thirty-dollar horse; but it was the only beast I could get hold of at the moment when a horse must be had. Besides, he turned out to be like a singed cat, a vast deal better than he looked.

I had repeatedly heard of a cow in the neighboring town, which was said to yield so much milk as to be the principal support of a small family whose head was a hopeless drunkard. She had cost seventy- five dollars, and had been a present to the drunkard's wife from one of her relatives. By careful inquiry, I satisfied myself that this cow gave twenty quarts daily, and that five months after calving, and on very indifferent pasture. I went to see her, and then her owner told me she was going to leave the place, and would sell the cow for fifty dollars. I did not hesitate a moment, but paid the money and had the cow brought home the same evening. My wife and daughter had not the least difficulty in learning to milk her. Under their treatment and my improved feeding, we kept her in full flow for a long time. She gave quite as much milk as two ordinary cows, while we had the expense of keeping only one. This I consider genuine good management: the best is always the cheapest.

The cow was never permitted to go out of the barnyard. A trough of water enabled her to drink as often as she needed, but her green food was brought to her regularly three times daily, with double allowance at night. I began by mowing all the little grass-plots about the house and lanes, for in these sheltered nooks the sod sends up a heavy growth far in advance of field or meadow. But this supply was soon exhausted, though it lasted more than a week: besides, these usually neglected nooks afforded several mowings during the season, and the repeated cuttings produced the additional advantage of maintaining the sod in beautiful condition, as well as getting rid of numberless weeds. When the grass had all been once mowed over, we resorted to the clover This also was mowed and taken to her; and by this treatment my little clover-field held out astonishingly. Long before I had gone over it once, the portion first mowed was up high enough to be mowed again. Indeed, we did secure some hay in addition. In this way both horse and cow were soiled. When the clover gave out, the green corn which I had sowed in rows was eighteen inches to two feet high, and in capital condition to cut and feed. It then took the place of clover. Both horse and cow devoured it with high relish. It was the extra sweet com now so extensively cultivated in New Jersey for market, and contained an excess of saccharine matter, which made it not only very palatable, but which sensibly stimulated the flow of milk.

The yield of green food which this description of corn gives to the acre, when thus sowed, is enormous. Not having weighed it, I cannot speak as to the exact quantity, but should judge it to be at least seven times that of the best grass or clover. Even without cutting up with a straw-knife, the pigs ate it with equal avidity. In addition to this, the cow was fed morning and night with a little bran. The unconsumed corn, after being dried where it grew, was cut and gathered for winter fodder, and when cut fine and mixed with turnips which had been passed through a slicer, kept the cow in excellent condition.

She of course got many an armful of cabbage-leaves during the autumn and all through the winter, with now and then a sprinkling of sliced pumpkins, from which the seeds had first been taken, as they are sure to diminish the flow of milk.

Thus I was obliged to lay out no money for either horse or cow, except the few dollars expended for bran. By this treatment I secured all the manure they made. By feeding the barnyard itself, as well as the hog-pen, with green weeds and whatever litter and trash could be gathered up, the end of the season found me with a huge manure pile, all nicely collected under a rough shed, out of reach of drenching rain, hot sun, and wasting winds. I certainly secured thrice as much in one season as had ever been made on that place in three. In addition to this, the family had had more milk than they could use, fresh, rich, and buttery. Even the pigs fell heir to an occasional bucket of skim-milk.

When our city friends came to spend a day or two with us, we were able to astonish them with a tumbler of thick cream, instead of the usual staple beverages of the tea-table. My wife evidently felt a sort of pride in making a display of this kind, and Kate invariably spread herself by taking our visitors to the barnyard, to let them see how expert she had become at milking. When they remarked, at table, on the surpassing richness of the cream, as well as the milk, my wife was very apt to reply –

"Yes, but when your turn comes to go in the country, be particular not to buy a cheap cow."

This remark generally led to inquiry, and then Kate was brought out with the whole story of our first and second cow, which she accordingly gave with illustrations infinitely more amusing than any I have been able to introduce. Indeed, her power of amplification sometimes astonished me. She told the story of our having been cheated by the old sinner, with such graphic liveliness, my wife now and then interposing a parenthesis, that the company invariably concluded it was by far the better policy to give a wide berth to cheap cows. I am not certain whether the fun occasioned by Kate's narratives was not really very cheaply purchased by the small loss we suffered on that occasion.

This abundance of milk wrought quite a change in our habits as to tea and coffee. At supper, during the summer, we drank milk only; but insensibly we ran on in the same way into cold weather. In the end, we found that we liked coffee in the morning only. This was a clear saving, besides being quite as wholesome. Our city milk bill had usually been a dollar a week. I am quite sure it did not cost over sixty cents a week to keep the cow. Then we had puddings and other dishes, which milk alone makes palatable, whenever we wanted them; and at any time of a hot summer's day a full draught of cold milk was always within reach. Then the quality was much superior, exceeding any thing to be found in city milk. I must admit that keeping a cow, like most other good things, involves some trouble; but my family would cheerfully undertake twice as much as they have ever had with ours, rather than dispense with this yet uncanonized saint of the barnyard.

CHAPTER XI.
A CLOUD OF WEEDS GREAT SALES OF PLANTS.

June came without my being obliged to hire any thing but occasional help on the farm. But when the month was fairly set in, I found every inch of my ploughed land in a fair way of being smothered by the weeds. I was amazed at the countless numbers which sprang up, as well as at the rapidity with which they grew. There was almost every variety of these pests. It seemed as if the whole township had concentrated its wealth of weeds upon my premises. In the quick, warm soil of New Jersey, they appear to have found a most congenial home, as they abound on every farm that I have seen. Cultivators appear to have abandoned all hope of eradicating them. Knowing that the last year's crop had gone to seed, I confess to looking for something of the kind, but I was wholly unprepared for the thick haze which everywhere covered the ground.

I can bear any quantity of snakes, but for weeds I have a sort of religious aversion. I tried one week to overcome them with the cultivator, but I made discouraging headway. I then bought a Regular horse-weeder, which cut them down rapidly and effectually. But meantime others were growing up in the rows, and corners, and by-places, where nothing but the hoe could reach them, and robbing the crops of their support. It would never do to cultivate weeds – they must be got rid of at any cost, or my crops would be worthless. Several neighboring farmers, who had doubtless counted on this state of things, came along about the time they supposed my hands would be full, looked over the fence at my courageous onslaught, laughed, and called out, "It's no use – you can't kill the weeds !" Such was the sympathy they afforded. If my house had been on fire, every one of them would have promptly hurried to the rescue; but to assist a man in killing his weeds was what no one dreamed of doing. He didn't kill his own.

In this dilemma I was forced to hire a young man to help me, contracting to give him twelve dollars a month and board him. He turned out sober and industrious. We went to work courageously on the weeds. I will admit that my man Dick was quite as certain as my neighbors that we could never get permanently ahead of them, and that thus lacking faith he took hold of the cultivator and weeder, while I attacked the enemy in the rows and by-places. I kept him constantly at it, and worked steadily myself. A week's labor left a most encouraging mark upon the ground. The hot sun wilted and dried up the weeds as we cut them off. Two weeks enabled us to get over the whole lot, making it look clean and nice, I congratulated myself on our success, and inquired of Dick if he didn't think we had got ahead of the enemy now. This was on a Saturday evening. Dick looked up at the sky, which was then black and showery, with a warm south wind blowing, and a broad laugh came over his features as he replied, "This will do till next time." The fellow was evidently unwilling either to encourage or to disappoint me.

That night a powerful rain fell, with a warm, sultry wind, being what farmers call "growing weather." I found it to be even so, good for weeds at least. Monday morning came with a hot, clear sun, and, under the combined stimulating power of sun, rain, and temperature, I found that in two nights a new generation had started into life, quite as numerous as that we had just overcome. As I walked over the ground in company with Dick, I was confounded at the sight. But I noticed that he expressed no astonishment whatever – it was just what he knew was to come – and so he declared it would be if we made the ground as clean as a parlor every week.

He said he never knew the weeds to be got out of Jersey ground, and protested that it couldn't be done. He admitted that they were nuisances, but so Avere mosquitoes. But as neither, in his opinion, did any great harm, so he thought it not worth while to spend much time or money in endeavoring to get rid of them. In either case he considered the attempt a vain one, and this was the whole extent of his philosophy. He had in fact been educated to believe in weeds. I was mortified at his indifference, for I had labored to infuse into his mind the same hatred of the tribe with which my wife and Kate had been so happily inoculated. But Dick was proof against inoculation – his system repudiated it.

But it set me to thinking. As to defining what a weed was, I did not undertake that, beyond pronouncing it to be a plant growing out of its proper place. Neither did I undertake to settle the question as to the endless variety there seemed to be or these pests, nor by what unaccountable agency they had become so thoroughly diffused over the earth. I could not fail to admit, however, that it seemed, in the providence of God, that whenever man ceased to till the ground and cover it with cultivated crops, at his almighty command there sprung up a profuse vegetation with which to clothe its nakedness. While man might be idle, it was impossible for nature to be so – the earth could not lie barren of every thing. But it seemed to me impossible that these ten acres of mine could contain an absolutely indefinite number of seeds of these unwelcome plants. There must be some limitation of the number. At what figure did it stop? Was it one million, or a hundred millions? Neither Dick nor myself could answer this question.

Yet I came resolutely to the conclusion that there must be a limitation, and that if we could induce all the seeds contained in the soil to vegetate, and then destroy the plants before they matured a new crop, we should ever afterwards be excused from such constant labor as we had gone through, and as was likely to be our experience in the future. I submitted this proposition to Dick – that if we killed all the weeds as they grew, the time would come when there would be no weeds to kill. It struck me as being so simple that even Dick, with all his doggedness, could neither fail to comprehend nor acknowledge it. He did manage to comprehend it, but as to acknowledging its force, one might have argued with him for a month. He utterly denied the premises – he had no faith in our Jersey weeds ever being killed, no matter how much luck we had thus far had with them, and I would see that he was right.

But having originated the dogma, I fully believed in it, and felt bound to maintain it; so Dick and I went resolutely to work a second time, as soon as the new crop was well out of the ground. The labor was certainly not as great as on the first crop, but it was hot work. I earned a file in my pocket, and kept my hoe as sharp as I have always kept my carving knife, and taught Dick to put his horse-weeder in prime order every evening when we had quit work. The perspiration ran in a stream from me in the hot sun, and a few blisters rose on my hands, but my appetite was rampant, and never have my slumbers been so undisturbed and peaceful.

About the third week in June we got through the second cleaning, and then rested. From that time to the end of the first week in July there had been no rain, with a powerfully hot sun. During this interval the weeds grew again, and entirely new generations, some few of the first varieties, but the remainder being new sorts. Thus there were wet-weather weeds and dry-weather weeds; and as I afterwards found, there was a regular succession of varieties from spring to winter, and even into December – cold-weather weeds as well as hot-weather weeds. Against each new army as it showed itself an onslaught was to be made. I was persuaded in my mind that the same army which we killed this year could not show itself the next, and that therefore there ought to be that number less. But Dick could not see this.

I observed, moreover, that each variety had its particular period when it vegetated, so that it might have time to get ahead and keep out of the way of its successor. It was evident that the seeds of any one kind did not all vegetate the same season. Herein was a wonderful provision of Providence to insure the perpetuity of all; for if all the rag-weed, for instance, had vegetated the first season of my experience, they would assuredly have been killed. But multitudes remained dormant in the earth, as if thus stored up for the purpose of repairing, another year, the casualties which their forerunners had encountered during the present one. Thus no one weed can be extirpated in a single season; neither do we have the whole catalogue to attack at the same time.

My warfare against the enemy continued unabated. As the time came for each new variety to show itself, so we took it in hand with hoe and weeder. Dick and his horse made such admirable progress, that I cannot refrain from recommending this most efficient tool to the notice of every cultivator. With one man and a horse it will do the work of six men, cutting off the weeds just below the ground and leaving them to wilt on the surface. It costs but six dollars, and can be had in all the cities. It would have cost me a hundred dollars to do the same amount of work with the hoe, which this implement did within four weeks.

Thus aided, our labors extended clear into November. In the intervals between the different growths f weeds, we looked after the other crops. But when the winter closed in upon us, the whole ground was so thoroughly cleaned of them as to be the admiration of the jeerers and croakers who, early in the season, had pitied my enthusiasm or ridiculed my anticipations. Even Dick was somewhat subdued and doubtful. I do not think a single weed escaped our notice, and went to seed that season.

I saw this year a beautiful illustration of the idea that there are specific manures for certain plants. I can hardly doubt that each has its specific favorite, and that if cultivators could discover what that favorite is, our crops might be indefinitely increased. On a piece of ground which had been sowed with turnips, on which guano had previously been sprinkled during a gentle rain, there sprang up the most marvellous growth of purslane that ever met one's eyes. The whole ground was covered with the rankest growth of this weed that could be imagined. Every turnip was smothered out. It seemed as if the dormant purslane-seed had been instantly called into life by the touch of the guano. It was singular, too, that we had noticed no purslane growing on that particular spot previous to the application of this rapidly-acting fertilizer.

I confess the sight of a dense carpet of purslane instead of a crop of turnips, almost staggered me as to the correctness of my theory that the number of seeds in the ground, yet to vegetate, must somewhere have a limit. Here were evidently millions of a kind which, up to this time, had not even showed themselves. After allowing the purslane to grow two weeks, Dick cut it off with his horse-weeder, raked it up; and carried it to the pigs, who consumed it with avidity. We then recultivated the ground and sowed again with turnips; but the yield was. very poor. Either the purslane had appropriated the whole energy of the guano, or the sowing was too late in the season.

But this little incident will illustrate the value of observation to a farmer. Book-farming is a good thing in its place, but observation is equally instructive. The former is not sufficient, of itself, to make good tillers of the soil. It will not answer in place of attentive observation. It forms, indeed, but the poorest kind of a substitute for that habit which every farmer should cultivate, of going all over his premises daily during the growing season, and noticing the peculiarities of particular plants; the habits of destructive animals or insects; the depredations as well as the services of birds; the when, the how, and the apparent wherefore of the germination of seeds; the growth of the stem, the vine, or the stalk that proceeds from them, and the formation, growth, and ripening of the fruit which they bear. Let no farmer, fruit-grower, or gardener, neglect observation for an exclusive reliance on book-farming.

It would be a most erroneous conclusion for the reader to suppose that all this long-continued labor in keeping the ground clear of weeds was so much labor thrown away. On the contrary, even apart from ridding the soil of so many nuisances, so many robbers of the nourishment provided for useful plants, it kept the land in the most admirable condition. The good conferred upon the garden by hoeing and raking, was re-enacted here. Every thing I had planted grew with surprising luxuriance. I do think it was an illustration of the value of thorough culture, made so manifest that no one could fail to observe it. It abundantly repaid me for all my watchfulness and care. Dick was forced to acknowledge that he had seen no such clean work done in that part of New Jersey.

My nurseryman came along at the end of the season, to see how I had fared, and walked deliberately over the ground with me, examining the peach- trees. He said he had never seen young trees grow more vigorously. Not one of them had died. The raspberries had not grown so much as he expected, but the strawberry-rows were now filled with plants. As runners were thrown out, I had carefully trained them in line with the parent stools, not permitting them to sprawl right and left over a great surface, forming a mass that could not be weeded, even by hand. This he did not approve of. He said by letting them spread out right and left the crop of fruit would be much greater, but admitted that the size of the berries would be much smaller. But he contended that *quantity* was what the public wanted, and that they did not care so much for *quality*. Yet he could not explain the damaging fact that the largest sized fruit was always the most eagerly sought after, and invariably commanded the highest price. Though he did not approve of my mode of cultivation, yet he could not convince me that I had made a mistake.

From these we walked over to the blackberries. They, too, had grown finely under my thorough culture of the ground. Besides sending up good canes which promised a fair crop the next season, each root had sent up several suckers, some of them several feet away, and out of the line of the row. These I had intended to sell, and had preserved as many as possible, knowing there would be a demand for all. The interest in the new berry had rapidly extended all round among my neighbors, and I very soon discovered that my nurseryman wanted to buy. In fact, I believe he came more for that purpose than to see how I was doing. But I talked offish – spoke of having engaged two or three lots, and could hardly speak with certainty. Finally, he offered to give me a receipt for the §120 he was to receive out of the strawberries he had sold me, and pay me §100 down, for a thousand blackberry plants. Though I felt pretty sure I could do better, yet I closed with him. As he had evidently come prepared with money to clinch some sort of bargain, he produced it and paid me on the spot. He afterwards retailed nearly all of the plants for a much larger sum. But it was a good bargain for both of us. It paid me well, and was all clear profit.

I may add that these blackberry roots came into more active demand from that time until the next spring; and when spring opened, more suckers came up, as if knowing they were wanted. These, with my previous stock, amounted to a large number. A seed man in the city advertised them for sale, and took retail orders for me. His sales, with my own, absorbed every root I could spare. When they had all been disposed of, and my receipts were footed up, I found that they amounted to four hundred and sixty dollars, leaving me three hundred and forty dollars clear, after paying for my strawberry plants.

This was far better than I had anticipated. It may sound curiously now, when the plants can be had so cheaply, but it is a true picture of the market at the time of which I write. It is the great profit to be realized from the sale of new plants that stimulates their cultivation. Many men have made fortunes from the sale of a new fruit or flower, and others are repeating the operation now. In fact, it is the hope of this great gain that has given to the world so many new and valuable plants, some originated from seed, some by hybridization, some from solitary hiding-places in the woods and mountains, and some by importation form distant countries. Success in one thing stimulates to exertion for another, and thus the race of a vast and intelligent competition is maintained. But the public is the greatest gainer after all.

My profits from this source, the first year, may by some be regarded as an exceptional thing, to be realized only by the fortunate few, and not to be regularly counted on. But this is not the case. There are thousands of cultivators who are constantly in the market as purchasers. If it were not so, the vast nursery establishments which exist all over the country could not be maintained. Every fruit-grower, like myself, has been compelled to buy in the beginning of his operations; but his turn for selling has invariably come round. As a general rule, whatever outlay a beginner makes in supplying himself with the smaller fruits, is afterwards reimbursed from the sale of surplus plants he does not need. This sale occurs annually, and in time will far exceed his original outlay.

If the plants be rare in the market, and if he should have gone into the propagation at a very early day, before prices have found their lowest level, his profits will be the larger. Hence the utmost watchfulness of the market should be maintained, New plants, better breeds of animals, and in fact every improvement connected with agriculture, if judiciously adopted at the earliest moment, will generally be found to pay, even after allowing for losses on the numerous cheats which are continually turning up.

CHAPTER XII.
PIGS AND POULTRY –
LUCK AND ILL LUCK.

Very early after taking possession, I invested twelve dollars in the purchase of seven pigs of the ordinary country breed. They were wanted to eat the many odds and ends which are yielded by ten acres, a good garden, and the kitchen. I did not look for much money profit from them, but I knew they were great as architects in building up a manure heap. Yet they were capital things with which to pack a meat-tub at Christmas, saving money from the butcher, as well as much running abroad to market. They shared with the cow in the abundant trimmings and surplus from the garden, eating many things which she rejected, and appropriating all the slop from the kitchen. In addition to this, we fed them twice a day with boiled bran, sometimes with a handful of corn meal, but never upon whole corn. This cooking of the food was no great trouble in the kitchen, but its effect on the pigs was most beneficial. They grew finely, except one which died after four months' feeding, but from what cause could not be ascertained.

The consequence was, that when October came round, the six remaining ones were estimated by Dick to average at least one hundred and fifty pounds each, and were in prime condition for fattening. In the early part of that month their supply of cooked mush was increased. I am of opinion that farmers leave the fattening of their hogs too late, and that a month on corn, before December, is worth three months after it. By the tenth of December they were ready for the butcher, and on being killed, were found to average two hundred and twenty-four pounds, or nine hundred and forty-four in all. This being three times as much as we needed for home use, the remainder was sent to the store, where it netted me forty-nine dollars.

I am quite certain there was a profit on these pigs. They consumed quantities of refuse tomatoes, and devoured parsnips with the greatest eagerness. One day I directed Dick to cut up some stalks of our green sweet corn, by means of the fodder-cutter, which delivers them in pieces half an inch long, and mix them with bran for the pigs. I found they consumed it with great avidity. Ever after that they were served twice daily with the same mess. It seemed to take the place of stronger food, as well as of grass, and was an acceptable variety. In this way the money cost of food was kept at a low figure, and the labor we spent on the pigs showed itself in the fine yield of prime pork, which brought the highest price in the market. The yield of rich manure was also very satisfactory, all which, at intervals through the season, was removed from the pen and put under cover, for manure thus housed from the sun and rain is worth about double that which is exposed all the year round. This was another item of profit: if the pigs had not manufactured it, money would have been required to pay for its equivalent.

After these six had been killed, I purchased seven others, some two months old, having abundance of roots, offal cabbages, and a stack of the sweet-corn fodder on hand. These seven cost the same as the others, twelve dollars. As Dick was found to be a good, trustworthy fellow, he was to be kept all the year round; and as he would be hanging about the barnyard during the winter, when the ground was wet and sloppy, looking after the horse and cow, the pigs would help to fill up his time. The cooking of food for both cow and pigs was a great novelty to him. At first he could not be made to believe in it. When I ventured to insinuate to him that it would be any thing but agreeable to him to eat his dinners raw, the force of the idea did not strike him. So much is there in the power of long-established habit. Yet he did condescend to admit that he knew all pigs throve better on plenty of common kitchen-swill than on almost any thing else. I told him there was but one reason for this, and that was because all such swill had been cooked. When the improvement made by the first lot of pigs became too manifest for even him to dispute, he, together with the pigs, acknowledged the com and gave in.

When out-door operations for the season were over, Dick undertook the whole business of cooking for the pigs and cow himself. In fact, on one occasion I succeeded in getting him to curry down both cow and pigs. They all looked and showed so much better for near a week thereafter, that coming on him unexpectedly one day, I found him repeating the operation of his own motion, and so he voluntarily continued the practice during the whole winter. The pigs seemed delighted with the process, and had very little scratching of their own to do. Their backs and sides were kept continually smooth, while their whole appearance was changed for the better. As to the cow, she took to being curried with the best possible grace, and improved under it as much as the pigs; but whether it increased the flow of milk I cannot say, as no means were taken to solve that question. But as Dick's devotion to the currycomb excited my admiration, so there was abundant evidence that both pigs and cow were equally captivated.

This business of raising and carefully attending to only half a dozen hogs, is worthy of every small farmer's serious study and attention. The hog and his food, with what is cheapest and best for him, is really one of the sciences, not an exact one, it is true, but still a science. One must look at and study many things, and they can all be made to pay. The propensity to acquire fat in many animals seems to have been implanted by nature. The hog fattens most rapidly in such a condition of the atmosphere as is most congenial to his comfort – not too hot, nor too cold. Hence the months of September, October, and November are the best for making pork. The more agreeable the weather, the less is the amount of food required to supply the waste of life. It has been found by some persons that a clover-field is the best and cheapest place to keep hogs in during the spring and summer months, where they have a plenty of water, the slop from the house, and the sour milk from the dairy. All sour feed contains more nitrogen than when fed in a sweet state. The first green herbage of the spring works off the impurities of the blood, cleanses the system, renovates the constitution, and enables the animal to accumulate a store of strength to carry it forward to its destined course.

Many object to beginning the fattening process so early in the season, as the corn relied on for that purpose is not then fully matured. But, taking all things into consideration, it is perhaps better to feed corn before it is ripe, as in that state it possesses more sweetness. Most varieties are in milk in September, when the hogs will chew it, swallow the juice, and eject the dry, fibrous matter. During the growing season of the year, swine can be fed on articles not readily marketable, as imperfect fruit, vegetables, &c. When such articles are used, cooking them is always economical. Most vegetables, when boiled or steamed, and mixed with only an eighth of their bulk of mill-feed or meal, whey, and milk left to sour, will fatten hogs fast. In this state they will eat it with avidity, and derive more benefit from it than when fed in an unfermented state. Articles of a perishable nature should be used first, to prevent waste, as it is desirable to turn all the products of the farm to the best account. Another quite important advantage of early feeding is the less trouble in cooking the food. Convenience of feeding is promoted, as there is no cost nor trouble to guard against freezing.

The more you can mix the food, the better, as they will thrive faster on mixed food than when fed separately. In feeding, no more should be given at a time than is eaten up clean, and the feeding should be regular as to time. It is of the greatest importance to get the best varieties, those that are well formed, and have an aptitude for taking on fat readily, and consume the least food. As to which is the best kind, there seems to be a great diversity of opinion, some preferring one kind and some another. The Suffolks come to maturity earliest, and probably are the most profitable to kill at from seven to ten months; but others prefer the Berk- shires. The pork of both is excellent: they will usually weigh from 250 to 300 pounds at the age of eight or ten months. The better way is to have the pigs dropped about the first of April, and feed well until December, and then butcher.

From a variety of experiments, I am satisfied it is wrong to let a hog remain poor twelve months of his life, when he could be made as large in nine months as he generally is in fifteen; and I conceive it a great error to feed corn to hogs without grinding. It has been proved by the Shakers, after thirty years' trial, that ground corn is one-third better for hogs and cattle-feed than if unground. In the case of another feeder, he ascertained the ratio of gain to be even greater than that of the Shakers. Others assert that cooking corn-meal nearly doubles its value. A distinguished agriculturist in Ohio proved that nineteen pounds of cooked meal were equal in value to fifty pounds raw. If pigs are well kept for three months after being dropped, they cannot be stunted after that, even if the supply of food is less than it should be.

It is desirable that hogs should be provided with a dry floor for eating and sleeping only, and the whole pen completely sheltered, to prevent any washing or waste of the manure. The commonwealth of the piggery should be furnished with plenty of straw, potato-vines, leaves, sawdust, and the like, with an occasional load of muck, and almost any quantity of weeds, all of which will be converted into the most efficient supports of vegetable life. Hogs are the best composters known, as they delight in upturning any such article as the farmer wishes to convert into manure for the coming year.

There can be no question as to its paying to make pork, though men differ on this as widely as their pork differs when brought to market. The poorer the pork, the more the owner complains of his profits, or rather of his losses; and the better the pork, the more is the owner satisfied. There can be no profit in raising a poor breed of hogs, that have no fattening qualities; nor even a good breed, without conveniences or proper care. A good hog cannot be fatted to any profit in mud or filth, nor where he suffers from cold. His comfort should be consulted as much as that of any other animal. It is a great error to assume that he is naturally fond of living among filth. On the contrary, hogs are remarkably neat, and those which fatten the best always keep themselves the cleanest. One farmer assured me that he had made his corn bring $1.25 per bushel by passing it through the bowels of his hogs, besides having the manure clear. Another did much better by cooking his meal.

As no farm is pronounced complete without poultry, and as both my wife and daughters were especially fond of looking after chickens, – at least they thought they would be, – so, to make their new home attractive, I invested $7 in the purchase of a cock and ten hens. They were warranted to be powerful layers, and would hatch fifteen eggs apiece. It struck me that this sounded very large, but on my wife observing it would be only a hundred and fifty chickens the first season, I gave in without a word. The fact is that chickens were not my hobby. I did not think they would pay, even after hearing my wife dilate on the luxury it would be to have fresh eggs every morning for breakfast, for pies and puddings, and various other things which she enumerated, and, as she expressed it, "eggs of our own laying."

I could not see how this circle of wonders was to be accomplished by only ten hens, and insinuated that it would be a good thing if she could make a bargain with each of her hens to lay two eggs a day. In reply to this, she astonished me by saying that Americans did not know how to make the most of things, but that the French did. She said that a certain Frenchman, mentioning his name – he was either a marquis or count, of course – had recently discovered the art of making hens lay every day by feeding them on horse-flesh, and that he feeds out twenty-five horses a day, which he obtains among the used-up hacks of Paris. She said he had a hennery which furnishes forty thousand dozens of eggs a week, and that it yields the proprietor a clear profit of five thousand dollars every seven days. After hearing this I felt certain she had been reading some modern poultry-book. But as she did not speak of requiring me to furnish horse-flesh for her pets, nor contemplate the establishment of a fresh-laid egg company, but only suggested the consumption of a little raw meat now and then, I volunteered no objections. Her enthusiasm was such as to make it unsafe to do so. Why should not she and the children be gratified?

The hens came home, and were put into a cage in the barnyard, to familiarize them with their new home. But they did not lay so freely as she had expected, while some did not lay at all. Worse than that, as soon as let out of their cage, they got over the fence into the garden, where they scratched as violently as if each one had a brood of fifteen to scratch for. They made terrible havoc among the young flowers and vegetables, and tore up the beds which had been so nicely raked. One of the girls was employed half her time in driving them out'. I thought it too great an expense to raise the barnyard fence high enough to keep them in, and so they were marched back into the cage. It happened to be too small for so many fowls, which my wife did not suspect, until one day, putting her hand in to draw forth a sick hen, she discovered her whole arm and sleeve to be swarming with lice. Here was something she did not remember to have been treated of in her poultry-book. But the nuisance was so great, as well as so active, soon extending itself all over her person, as to compel her to strip and change her entire dress, and to plunge the lousy one in a tub of -water.

I confess the difficulty was a new one to me. My experience in poultry had been limited. My knowledge of them was exclusively anatomical, obtained by frequent dissections with the carving-knife. On calling Dick, however, it appeared that he knew more about this trouble than the whole family together. When my wife described her condition to him, and how she had swarmed with the vermin, the fellow laughed outright, but said they wouldn't hurt – he knew all about them, for he had been full of lice more than once ! He said he expected this, as the fowls had been kept up too close: they would neither lay, thrive, nor keep clear of vermin, unless allowed to run about.

But he took the case in hand, clipped their wings, saturated their heads with lamp oil, provided abundance of ashes for them to roll in, and then turned them loose in the barnyard. He then obtained poles of sassafras wood for them to roost on, as he said the peculiar odor of that tree would drive the enemy away. I presume his prescriptions answered the purpose; at all events, we discovered no more hen-lice, because the whole family were careful never to touch a fowl again.

I think this little catastrophe took all the romance out of my wife touching chickens. I rarely heard her mention eggs afterwards, except when some of us were going to the store for other things, and she was careful never to purchase chickens with the feathers on. She never referred to the hundred and fifty she was to hatch out that season; nor have I ever heard her even mention horse-flesh as a sure thing for making hens lay all the year round. That winter Dick fattened and killed the whole lot. My wife did not seem to have much stomach for them when they came upon the table. I was not sorry for it, except that she had been disappointed. Her knowledge of keeping poultry had been purely theoretical, and her first disappointment had completely weaned her of her fondness for the art.

But this brief and unlucky experience of ours should by no means operate to discourage others. Money is undoubtedly made by skillful men at raising poultry. It cannot be a losing business, or so many thousand tons would not be annually produced. Volumes have been written on the subject, which all who contemplate embarking in the business may consult with profit. As an incident of farm life it will always be interesting, and with those who understand the art it ought to be profitable.

Foreigners must be more experienced in the business of raising poultry than Americans, judging by the vast quantities they annually produce for market. The quantity imported into England is so enormous, that it is impossible to determine its amount. Into only two of the principal London markets there is annually brought from France and Belgium, 75,000,000 eggs, 2,000,000 fowls, 400,000 pigeons, 200,000 geese and turkeys, and 300,000 ducks. In addition to these, the large amount sent to poulterers and private houses must be considered. The Brighton railroad alone carries yearly 2,600 tons of eggs which come from France and Belgium. Yet, with all these immense supplies, the London markets are frequently very meagrely supplied with butter and eggs. The trade is shown by these figures to be one of great national value. Americans have strangely neglected its cultivation with the method and precision of foreigners. We can raise food more cheaply than they, while none of them can boast of possessing our incomparable Indian corn.

There are several of my neighbors who are highly skilled in the art of raising poultry. One of them is quite a poultry-fancier, and, by keeping only choice breeds, he realizes fancy prices for them. Another confines his fowls in a plum-orchard, and thus secures an annual crop of plums without being stung by the curculio. In general, the female portion of the family attend to this branch of domestic business, and realize a snug sum from it annually. A brood of young chickens turned into a garden, the hen confined in her coop, will soon clear it of destructive insects.

CHAPTER XIII.
CITY AND COUNTRY LIFE CONTRASTED.

THE pensive reader must not take it for granted that in going into the country we escaped all the annoyances of domestic life peculiar to the city, or that we fell heir to no new ones, such as we had never before experienced. He must remember that this is a world of compensations, and that nowhere will he be likely to find either an unmixed good or an unmixed evil. Such was exactly our experience. But on summing up the two, the balance was decidedly in our favor. It is true that though the town close by us had well-paved streets, yet the walk of half a mile to reach them was a mere gravel path, which was sometimes muddy in summer, and sloppy with unshovelled snow in winter. But I walked over it almost daily to the post- office, not even imagining that it was worse than a city pavement. The tramp of the children to school was not longer than they had been used to, and my wife and daughters thought it no hardship to go shopping among the well-supplied stores quite as frequently as when living in the city. Indeed, I sometimes thought they went a little oftener. They were certainly as well posted up as to the new fashions as they had ever been, while the fresh country air, united with constant exercise, kept them in good appetite, even to the rounding of their cheeks, and the maintenance of a better color in them than ever.

As to society, they very soon made acquaintances quite as agreeable as could be desired. Visiting became a very frequent thing; and after a few months I let in a suspicion that the girls found twice as many beaux as in the city, though there the average number is always larger than in the country. On throwing out an insinuation of this kind to Kate, one summer evening, after a large party of young folks had concluded their visit, she made open confession that it was so, and volunteered her conviction that they were decidedly more agreeable. I admit this confession did not surprise me, as there was

one young man among the party who had become especially attentive to Kate – bringing her the new magazines as soon as they were
out, sundry books and pictorials, and always having a deal to say to
her, with a singular genius for getting her away from the rest of the
company, so that most of their mysterious small-talk could be heard
by none but themselves. Another remark which I made to Kate on a
subsequent occasion, touching this subject, covered her bright face
with so many blushes that I ventured to mention the whole matter
to my wife; but she made so light of the thing that I said no more at
the time, thinking, perhaps, that the women were most likely the best
judges in such cases. But I have since discovered that my prognosti-
cations were much more to be depended on than hers.

Then the walks for miles around us were excellent, and we all became great walkers, for walking we found to be good. Not merely step-
ping from shop to shop, or from neighbor to neighbor, but stretching
away out into the country, to the freshest fields, the shadiest woods,
the highest ridges, and the greenest lawns. We found that however
sullen the imagination may have been among its griefs at home, here
it cheered up and smiled. However listless the limbs may have been
by steady toil, here they were braeed up, and the lagging gait became
buoyant again. However stubborn the memory may be in presenting
that only which was agonizing, and insisting on that which cannot
be retrieved, on walking among the glowing fields it ceases to regard
the former, and forgets the latter. Indeed, we all came to esteem the
mere breathing of the fresh wind upon the commonest highway to
be rest and comfort, which must be felt to be believed.

But then we had neither gas nor hydrant water, those two prime
luxuries of city life. Yet there was a pump in a deep well under a shed at
the kitchen door, from which we drew water so cold as not at any time
to need that other city luxury, ice. It was gratifying to see how expertly
even the small children operated with the pump-handle. In a month
we ceased to regret the hydrants. As to gas, wo had the modern lamps,
which give so clear a light; not so convenient, it must be confessed,
but then they did not cost us over half as much, neither did we sit up

near so long at night. There were two mails from the city daily, and the newsboy threw the morning paper into the front door while we sat at breakfast. The evening paper came up from the city before we had supped. We had two daily mails from New York, besides a telegraph station. The baker served us twice a day with bread, when we needed it; the oysterman became a bore, he rang the bell so often; and the fish-wagon, with sea-fish packed in ice, directly from the shore, was within call as often as we desired, with fish as cheap and sound as any to be purchased in the city. Groceries and provisions from the stores cost no more than they did there, but they were no cheaper. But in the item of rent the saving was enormous, – really half enough, in my case, to keep a moderate family. Many's the time, when sweating over the weeds, have I thought of this last heavy drain on the purse of the city toiler, and thanked Heaven that I had ceased to work for the landlord.

We had books as abundantly as aforetime, as we retained our share in the city library, and became subscribers to that in the adjoining town. It is true that the road in front of us was never thronged like Chestnut-street, but we neither sighed after the crowd nor missed its presence. We saw no flash of jewelry, nor heard the rustling of expensive silks, except the few which on particular occasions were sported within our own unostentatious domicil. Our entire wardrobes were manifestly on a scale less costly than ever. Our old city friends were apparently a great way off, but as they could reach us in an hour either by steamboat or rail, they quickly found us out. The relish of their society was heightened by distance and separation. In short, while far from being hermits, we were happy in ourselves. I think my wife became a perfectly happy woman – what it had been the great study of my wedded life to make her – the very sparkle and sunshine of the house. She possessed the magic secret of being contented under any circumstances. The current of my life had never been so dark and unpropitious, that the sunshine of her happy face, falling across its turbid course, failed to awake an answering gleam.

Speaking of visitors from abroad, I noticed that our city friends came to make their visits on the very hottest summer days, when, of all others, we were ourselves sufficiently exhausted by the heat, and were disposed to put up with as little cooking and in-door work as possible. But as such visitations Avere not exactly comfortable to the visited, so we could not see how they could be any more agreeable to the visitors. Yet they generally remarked, even when the mercury was up to ninety-five, "How much cooler it is in the country !" They did really enjoy either themselves or the heat. But my wife told them it was only the change of scene that made the weather tolerable, and that if they lived in the country they would soon discover it to be quite as hot as in the city. For my part, I bore the heat admirably, though tanned by the sun to the color of an aborigine; but I enjoyed the inexpressible luxury of going constantly in my shirt sleeves. I can hardly find words to describe the feeling of comfort which I enjoyed for full seven months out of the twelve from this little piece of latitudinarianism, the privilege of country life, but an unknown luxury in the city.

I saw that this press of company in the very hottest weather imposed an unpleasant burden on my wife, for she and my two oldest daughters were the sole caterers; and I intended to say something to her concerning it, as soon as a large party, then staying several days with us, should have concluded their visit. But on going into our chamber that very evening, she surprised me by asking if I could tell her why, when Eve was made from one of Adam's ribs, there was not a hired girl made at the same time, for to her mind it took three to make a pair – he, she, and a hired girl. I replied that I had not given much time to the study of navigation, but that I quite understood her meaning, and that it was exactly what I had myself been thinking of. If Adam's rib, after producing Eve, had not held out to produce a hired girl also, I told her there was a much quicker way of getting what she wanted, and that the first morning paper she might pick up would produce her twenty hired girls.

In this way, before the summer was over, I procured her a servant, thus making her little establishment complete. For this luxury we paid city wages. But this was a small item, when I saw how much her presence relieved my wife. After that, I do not think she complained quite as much of the hot weather, nor was she inclined so frequently to repeat her former observation, that the sultry days always brought the most company. Indeed, I am certain that on one or two occasions, when the dog-days were terribly oppressive, she prevailed on different parties to prolong their stay for nearly a week.

Now, this taking on of Betty did not imply that my daughters were to be brought up to do nothing – or to do every thing that is fashionable imperfectly. My wife had already educated them in domestic duties – not merely to marry, to go off *with* husbands in a hurry, and afterwards *from* them. To the two eldest she had taught a trade, and they were both able to earn their salt. They could not only dress themselves, but knew how to make their dresses and bonnets, and all the clothing for the younger children. She cultivated in them all that was necessary in the position in which they were born, one thing at a time, but that thing in perfection; so that if parents were impoverished, or if in after-life reverses should overtake themselves, they might feel independent in the ability to earn their own support. She frowned upon the senseless rivalries of social life, as destructive of morals, mind, and health, and imbued their spirits with a devout veneration for holy things. She taught them no worship of the almighty dollar, but sound, practical economy, the art of saving the pieces. Surely it must be education alone which fills the world with two kinds of girls – one kind which appears best abroad, good for parties, rides, and visits, and whose chief delight is in such things – good, in fact, for little else. The other is the kind that appears best at home, graceful in the parlor, captivating in social intercourse, useful in the sick- chamber as in the dining-room, and cheerful in all the precincts of home. They differ widely in character. One is often the family torment; the other the family blessing: one a moth consuming every thing about her; the other a sunbeam, inspiring life and gladness all along her pathway. As my wife embodied in herself all that to me appeared desirable in woman, so she possessed the faculty of transfusing her own virtues into the constitution of her daughters.

CHAPTER XIV.
TWO ACRES IN TRUCK –
REVOLUTION IN AGRICULTURE.

I HAD one acre in tomatoes, a vegetable for whose production the soil of New Jersey is perhaps without a rival. The plants are started in hot-beds, where they flourish until all danger from frost disappears, when they are set out in the open air, with a generous shovel-full of well-rotted stable manure deposited under each plant. A moist day is preferred for this operation; but even without it this plant generally goes on growing. It has been observed that the oftener it is trans-planted, the more quickly it matures; and as the great effort among growers is to be first in market, so some of them take pains to give it two transplantings. Having no hot-bed on my premises, and my time being fully occupied with other things, I was compelled to purchase plants from those who had them to spare, the cost of which is else-where stated. But the operation paid well.

The quantity produced by an acre of well-manured tomatoes is almost incredible. When in full bearing, the field seems to be per-fectly red with them. Those which come first into market, even with-out being perfectly ripe, sell for sixpence apiece. So popular has this vegetable become, and so great is the profit realized by cultivating it, that for nearly twenty years it has been grown in large quantities by Jerseymen who emigrated to Virginia for the purpose of taking advantage of the earlier climate of that genial region. There they bought farms, improved them by using freely the unappropriated and unvalued stores of manure to be found in the vicinity, and pro-duced whole cargoes of the choicest early vegetables required by the great consuming public of the northern cities. They shipped them hither two weeks ahead of all the Jersey truckers, and were rewarded by fabulous prices, from the receipt of which large fortunes result-ed. This mutually advantageous traffic had become a very important one, when rebellion broke it up. Intercourse was stopped, cultivation was abandoned, and the Virginia truckers were mined.

Although this competition seriously interfered with the profits of New Jersey farmers, yet it did not destroy them. The cultivation of early track and fruit continued to pay, though not so well as formerly. When prices fell, the Southern growers could not afford the cost of delivery here, and thus left us in undisputed possession of the market. But, as a general rule, the Virginia competitors invariably obtained the highest prices. A great portion of their several crops, however, perished on their hands; because, as they had no market here when prices fell, so the scanty population around them afforded none at home.

For the first few baskets of early tomatoes I sent to market, I obtained two dollars per basket of three pecks each. Other growers coming in competition with me, the price rapidly diminished as the supply increased, until it fell to twenty-five cents a bushel. At less than this the growers refused to pick them; and seasons have been repeatedly known when tens of thousands of bushels were left to perish on the vines. When this low price could be no longer obtained, they were gathered and thrown to the pigs, who consumed them freely. But as the season advanced the supply diminished, and the price again rose to a dollar a basket, the demand continuing as long as any could be procured. The tomatoes are at this season picked green from the vines, and placed under glass, where they are imperfectly ripened; but such is the public appreciation of this wholesome vegetable, that when thus only half reddened, they are eagerly sought after by hotels and boardinghouses.

But of latter years measures have been taken to prevent, to some extent, the enormous waste of tomatoes during the height of the season, by preserving them in cans. Establishments have been started, at which any quantity that may be offered is purchased at twenty-five cents a bushel; and now they can be kept through the whole year, and be preserved for winter consumption, the same as potatoes or turnips. By hermetically sealing them in cans from which the air has been expelled by heat, they are not only preserved, but made to retain their full flavor; and may be enjoyed, at a very moderate cost, in the winter as well as the summer. The demand for them is constant, large, and increasing, and putting up canned tomatoes has become an extensive business. One person, who commenced the business two years ago, is literally up to the eyes in tomatoes once a year. He provides for a single year's trade over fifty thousand cans, all of which are manufactured by himself; and he employs over thirty persons, most of them women. He engages tomatoes at twenty-five cents a bushel, a price at which the cultivator clears about a hundred dollars an acre, and they come in at the rate of a hundred and fifty bushels a day, requiring the constant labor of all hands into the night to dispose of them.

The building in which the business is carried on was constructed expressly for it. At one end of the room in which the canning is done is a range of brick-work supporting three large boilers; and adjoining is another large boiler, in which the scalding is done. The tomatoes are first thrown into this sealder, and after remaining there a sufficient time, are thrown upon a long table, on each side of which are ten or twelve young women, who rapidly divest them of their leathery hides. The peeled tomatoes are then thrown into the boilers, where they remain until they are raised to a boiling heat, when they are rapidly poured into the cans, and these are carried to the tinmen, who, with a dexterity truly marvellous, place the caps upon them, and solder them down, when they are piled up to cool, after which they are labelled, and are ready for market. The rapidity and the system with which all this is done is most remarkable, one of the tinmen soldering nearly a hundred cans in an hour.

The tomatoes thus preserved are readily salable in all the great cities, both for home consumption and for use at sea. Thus, few vegetables have gained so rapid and wide-spread a popularity as this. Until lately, but few persons would even taste them; and they were raised, when cultivated at all, more from curiosity than any thing else. Now, scarcely a person can be found who is not fond or them, and they occupy a prominent place on almost every table.

My single acre of tomatoes produced me a clear profit of §120. I am aware that others have realized more than double this amount, but they were experienced hands at the business. My gains were quite as much as I had anticipated.

From all the remainder of the three acres but little money was produced. It gave me parsnips, turnips, and pumpkins. Between the rows of sweet corn a fine crop of cabbages was raised, of which my sales amounted to §82. Thus, an abundant supply of succulent food was provided for horse, cow, and pigs during the winter, all which saved the outlay of so much cash. I admit that a few of my vegetables did not yield equal to the grounds of some of my neighbors, thus disappointing some of my calculations. But I was inexperienced, had much to learn, and was not discouraged. On the other hand, I had gone far ahead of them in the growth of my standard fruits; and the evident hit I had made with the new blackberry had the effect of impressing them with considerable respect for my courage and sagacity.

This business of raising vegetables for the great city markets, "trucking," as it is popularly called, is now the great staple of New Jersey agriculture. All the region of country stretching from Camden some forty miles towards New York, once enjoyed the reputation of being either all sand or all pine. It is traversed by the old highway between Philadelphia and New York, laid out by direction of royalty in colonial days, and protected at various points by barracks, in which troops were garrisoned. Some of the barracks remain to this day; though in chambers where high military revel once was held, devout congregations now worship. Along this royal highway passed all the early travel between the colonies; and after they had been severed from their parent stem, up to the advent of steamboats and railroads, it was the only thoroughfare between the two cities of New York and Philadelphia. Stages occupied five weary days between them, the horses exhausted by wading through a deep, laborious sand in summer, or the still deeper mud through which they floundered in winter. On miles of this road the sand was frightful. No local authorities worked it, no merciful builder of turnpikes ever thought of reclaiming it. It lay from generation to generation, as waste and wild as when the native pines were first cleared away. Access was so laborious, that few strangers visited the region through which it passed; and the land was held in large tracts, whereon but few settlers had made clearings. AU judged the soil as worthless as the deep sand in the highway. Where some settler did clear up a farm, his labors presented no inviting spectacle to the passing traveller. If manure was known in those days, the farmer did not appear to value it, for he neither manufactured nor used it. Phosphates and fertilizers had not been dreamed of. If he spread any fertilizer over his fields, it was but a starveling ration; hence his corn crop was a harvest of nubbins. Wheat he never thought of raising; rye was the sole winter grain, and rye bread, rye mush, and rye pie-crust, held uncontested dominion, squalid condiments as they all are, in each equally squalid farm-house. Ragweed and pigweed took alternate possession of the fields; cultivation was at its last point of attenuation; none grew rich, while all became poor; and as autumn came on, even the ordinarily thoughtless grasshopper climbed feebly up to the abounding mullein top, and with tears in his eyes surveyed the melancholy desolation around him. Such is a true picture of the king's highway up to the building of the Camden and Amboy Railroad.

No wonder that the great public who traversed it through this part of New Jersey should think it, and speak of it everywhere, as being all sand, seeing that in their passage through it they beheld but little else. Hence, the reputation thus early established continues to the present day, and the tradition has been incorporated into the public vernacular. The sandy road alone was seen, while the green and fertile tracts that lay beyond and around it were unknown, because unseen. Like the traveller from Dan to Beersheba, the cry was that all was barren. But time, improvement, education, railroads, and the marvellous growth of Philadelphia, New York, and fifty intermediate towns, have changed all this as by enchantment. Every mile of the old highway is now a splendid gravel turnpike, intersected by a dozen similar roads, which stretch away up into the country.

As good roads invite settlement, so population, the great pro-
moter of the value of land, has come in rapidly, and changed the
aspect of every farm-house. Good fences line the roadside, rank
hedge-rows have disappeared, new farm houses have been everywhere built, low lands have been drained, manures have been imported from the cities, wheat is now the staple winter-grain, rye has
ceased to be cultivated, and rye bread is now a mere reminiscence of
the old dispensation. But chief, perhaps, of all, the whole agricultural
world of New Jersey has been educated by the agricultural press to
a high standard of intelligence and enterprise. Its labors have led to
the establishment of numerous extensive nurseries, by the pressure
of a general demand for trees and smaller fruits, whose wilderness
of blossoms now annually blush and brighten upon every farm. It
has taught them to cultivate new vegetables and fruits for city consumption alone, salable for cash in each successive month; in doing
which, they have changed from a poverty-stricken to a money-making generation. It has taught them, what none previously believed,
that no good farming can be done without high manuring, and banished the ignorance and meanness that prevented them from spending money to secure it. It has introduced to their notice new and
portable manures, improved tools, better breeds of stock of all kinds,
and sharpened their perceptions, until they have now become men
of business as well as farmers, and so proved its value to them, that
he upon whose table no agricultural journal can be found, may be
written down as the laggard of a progressive age.

But in addition to all these stimulants to progress the Camden
and Amboy Railroad came in, giving it a vast momentum. Terminat-
ing at Philadelphia and New York, it opened up a cash market among
thousands asking for daily bread. When this road was first opened,
its annual way-freight yielded less than one hundred dollars a year.
But its managers wisely built station-houses at every cross-road, as
the farmers called for them. To these nuclei the produce of entire
townships quickly gathered in astonishing quantities. Agents from
the great cities traversed the country, and bought every thing that

was for sale. A cash market being brought to their very doors, where none had previously existed, an immense stimulus to production followed, and a new spirit was infused into the whole region. Hundreds of farms were renovated, cleared of foul weeds, drained, and liberally manured. New vegetables were cultivated. Tomatoes, peas, rhubarb, and early potatoes rose into prime staples. Green corn has been taken from a single county to the extent of two thousand tons daily. Other products go to market by thousands of baskets at a time. Way-trains are run for the sole accommodation of this truck business, stopping every few miles to take in the waiting contributions collected at the stations. To both railroad and farmer it has proved a highly remunerating traffic. These way-freights, thus wisely cultivated by the railroad, now amount to many thousands annually, and are steadily growing larger. Meantime, steamboats on the Delaware stop several times daily at new wharves on the river, sometimes taking at one trip two thousand baskets of truck, from a point where, twelve years ago, the same number could not be gathered during an entire season. The grower thus has the choice of the two richest markets in the country. He reaches Philadelphia in one hour, and New York in three.

It must be manifest that crops of such magnitude cannot be produced on mere sand. Hence the traditional notion that New Jersey is a sand heap, desolate and barren at that, has long been proved to be a fallacy. Men do not grow rich upon a burning desert, such as this region has been described. Yet the farmers who occupy it are notoriously becoming so. They lend money annually on mortgage, after spending thousands in manure, while farms have advanced from §30 to §100 and §200 per acre. The last ten years have added thirty per cent, to the population. Schools, churches, and towns have proportionately increased in number.

The soil of this truck region contains a large proportion of sand with loam, on which manure acts with an energetic quickness that brings all early truck into the great markets in advance of the neighboring country. This secures high prices. Southern competition has only stimulated the growers to increased exertion. Though from this cause losing some of the high rewards of former years, yet the aggregate of profit does not seem to diminish. Better cultivation, higher manuring, changing one product for another, with more land brought into tillage, enable them to foot up as large an amount of sales at the end of the season as aforetime. They see that the world cannot be overfed, and that any thing they can produce will command a ready market. Consumers increase annually, and the public appetite loses none of its rampant fierceness. Hence, competition stimulates instead of discouraging.

A vast area is planted with tomatoes. Though thousands of bushels perish every season, yet two hundred, and even four hundred dollars an acre is frequently the clear profit. Thirty years ago, three bunches of rhubarb were brought to the London market for sale, but as no one could be found to buy them, they were given away; yet London now consumes seven thousand tons annually. So, in New Jersey – the planter of the first half acre was pitied for his temerity. Now, there are hundreds of acres of rhubarb. The production of peas, pickles, cucumbers, melons, and cabbages is immense. Early com is raised in vast quantities. All these various products command cash on delivery.

The soil of this region has long been famous for its growth of melons. Formerly they were raised by ship-loads, but Southern competition has checked their production. Yet New Jersey citrons possess a flavor so exquisite, that they cannot be driven from the market. Peaches have long since become almost obsolete, the yellows and the worm having been great discouragements. But within three years, hundreds of acres of them have been planted in New Jersey, and the nurseries find ready sale, in seasons of average prosperity, for all they can produce. Numerous orchards will annually come into bearing; and the chances are that this once famous staple will again be domesticated in its ancient stronghold. Among the smaller fruits, strawberries occupy an important place in New Jersey, whose soil seems peculiarly adapted to them. The yield per acre is enormous. One grower has gathered 400 bushels from three acres of the Albany seedling. He began his plantation with a single dozen plants, at $2.50 per dozen. New

York and Philadelphia took them all at an average of eighteen cents a quart. This patch was a marvel to look at. The ground appeared fairly red with berries of great size, and were so abundant that pickers abandoned other fields at two cents a quart, and volunteered to pick this at one and a half. Other neighboring growers realized large returns. The two counties of Burlington and Monmouth are believed to yield more berries of all kinds than any district of equal area in the Union, and the cultivation is rapidly extending.

A year or two ago, somebody invented and patented a new box for taking them to market, lighter, neater, cheaper than the old one, and securing thorough ventilation to the fruit. A club of Connecticut men forthwith organized a company with a capital of $10,000 for manufacturing them; built a factory, started an engine, and now have forty hands at work. An agent of the company went through the State last fall, from Middletown to Camden, showing samples, and taking orders. He sold three hundred thousand boxes, many to those who had the old ones, but more to others just wanting them. As he travelled on foot, with samples in his hand, he inquired his way over the country, from farm to farm, and probably discovered every grower of an acre of berries. Of course he could not fail to visit and supply me. He gave me many curious items of information touching the extent of the berry business. There are parties in this country who have fifty acres of strawberries on a single farm, with a thousand dollars invested merely in the small boxes in which they are taken to market. He reports that the two counties of Burlington and Monmouth produce more berries than all the remainder of the State. Strawberries and raspberries are now the staples, to which the blackberry has recently been added. The great consuming stomach of the large cities, having long been fed on these delicious fruits, must continue to buy. Growers seem to know that after thirty years' propagation of the strawberry, this devouring stomach has never been surfeited, – that the more it is fed the more it consumes.

CHAPTER XV.
BIRDS, AND THE SERVICES
THEY RENDER.

ONE morning in September, hearing shots fired repeatedly at the farther end of my grounds, and proceeding thither to ascertain the cause, I discovered three great, overgrown boobies, with guns in their hands, trampling down my strawberries, and shooting bluebirds and robins. On inquiring where they belonged, they answered in the next township. I suggested to them that I thought their own township was quite large enough to keep its own loafers, without sending them to depredate on me, warned them never to show themselves on my premises again, and then drove them out. This happened to be the only occasion on which I was invaded by any of the worthless, loafing tribe of gunners, who roam over some neighborhoods, engaged in the manly occupation of killing tomtits and catbirds.

For all such my aversion was as decided as my partiality for the birds was strong. One of the little amusements I indulged in immediately on taking possession of my farm, was to put up at least twenty little rough contrivances about the premises, in which the birds might build. Knowing their value as destroyers of insects, I was determined to protect them; and thus, around the dwelling-house, in the garden-trees, and upon the sides of the barn, as well as in other places which promised to be popular, I placed boxes, calabashes, and squashes for them to occupy. The wrens and bluebirds took to them with gratifying readiness, built, and reared their families. But I observed that the wren quickly took possession of every one in which the hole was just large enough to admit himself, and too small to allow the bluebird to enter; while in those large enough to admit a bluebird no wren would build. This was because the bluebird has a standing spite against the wrens, which leads him to enter the nests of the latter, whenever possible, and destroy their eggs. Almost any number of wrens may thus be attracted round the house and garden, where they act as vigilant destroyers of insects.

These interesting creatures soon hatched out large broods of young, to provide food for which they were incessantly on the wing. They became surprisingly tame and familiar, those especially which were nearest the house, and in trees beneath which the family were constantly passing. We watched their movements through the season with increasing interest. No cat was permitted even to approach their nests, no tree on which a family was domiciled was ever jarred or shaken; and the young children, instead of regarding them as game to be frightened off, or hunted, caught, and killed, were educated to admire and love them. Indeed, so carefully did we observe their looks and motions, that many times I felt almost sure that I could identify and recognize the tenants of particular boxes. They ranged over the whole extent of my ten acres, clearing the bushes and vegetables of insects and worms; while the garden, in which they sang and chattered from daybreak until sunset, was kept entirely clear of the destroyers. I encountered them at the furthest extremity of my domain, peering under the peach-leaves, flitting from one tomato-vine to another, almost as tame as those at home. They must have known me, and felt safe from harm. I am persuaded that I recognized them. Yet it was at this class of useful birds that the boobies calling themselves sportsmen were aiming their weapons, when I routed them from the premises, and forbid the murderous foray.

Insects are, occasionally, one of the farmer's greatest pests. But high, thorough farming is a potent destroyer. It is claimed by British writers to be a sure one. When the average produce of wheat in England was only twenty bushels per acre, the ravages of the insect tribe were far more general and destructive than they have been since the average has risen to forty bushels per acre. Why may not the cultivation of domestic birds like these, that nestle round the house and garden, where insects mostly congregate, be considered an important feature in any system of thorough farming?

Besides the wrens and bluebirds, the robins built under the eaves of the wood-shed, and became exceedingly tame. The more social swallow took possession of every convenient nestling-place about the barn, while troops of little sparrows came confidingly to the kitchen door to pick up the crumbs of bread which the children scattered on the pavement as soon as they discovered that these innocent little creatures were fond of them. Thus my premises became a sort of open aviary, in which a multitude of birds were cultivated with assiduous care, and where they shall be even more assiduously domesticated, as long as I continue to be lord of the manor. I pity the man who can look on these things, who can listen to the song of wrens, the loud, inspiring carol of the robin on the tree-top, as the setting sun gilds its utmost extremities, listening to these vocal evidences of animal comfort and enjoyment, without feeling any augmentation of his own pleasures, and that the lonesome blank which sometimes hangs around a rural residence is thus gratefully filled.

One morning, hearing a great clamor and turmoil in a thicket in the garden, where a nest of orioles had been filled with young birds, I cautiously approached to discover the cause. A dozen orioles were hovering about in great excitement, and for some time it was impossible to discover the meaning of the trouble. But remaining perfectly quiet, so as not to increase the disturbance, I at length discovered an oriole, whose wing had become so entangled in one end of a long string which formed part of the nest that she could not escape. The other birds had also discovered her condition, and hence their lamentation over a misfortune they were unable to remedy. But they did all they could, and were assiduously bringing food to a nest full of voracious young ones, as well as feeding the imprisoned parent. I was so struck with the interesting spectacle that my family were called out to witness it; then, having gazed upon it a few moments, I cautiously approached the prisoner, took her in my hands, carefully untied and then cut away the treacherous string, and let the frightened warbler go free. She instantly flew up into her nest, as if to see that all her callow brood were safe, gave us a song of thanks, and immediately the crowd of sympathizing birds, as if conscious that the difficulty no longer existed, flew away to their respective nests.

It takes mankind a great while to learn the ways of Providence, and to understand that things are better contrived for him than he can contrive them for himself. Of late, the people are beginning to learn that they have mistaken the character of most of the little birds, and have not understood the object of the Almighty in creating them. They are the friends of those who plant, and sow, and reap. It has been seen that they live mostly on insects, which are among the worst enemies of the agriculturist; and that if they take now and then a grain of wheat, a grape, a cherry, or a strawberry, they levy but a small tax for the immense services rendered. In this altered state of things, legislatures are passing laws for the protection of little birds, and increasing the penalties to be enforced upon the bird-killers.

A farmer in my neighborhood came one day to borrow a gun for the purpose of killing some yellow-birds in his field of wheat, which he said were eating up the grain. I declined to loan the gun. In order, however, to gratify his curiosity, I shot one of them, opened its crop, and found in it two hundred weevils, and but four grains of wheat, and in these four grains the weevil had burrowed! This was a most instructive lesson, and worth the life of the poor bird, valuable as it was. This bird resembles the canary, and sings finely. One fact like this affords an eloquent text for sermonizing, for the benefit of the farmers and others who may look upon little birds as inimical to their interests. Every hunter and farmer ought to know that there is hardly a bird that flies that is not a friend of the farmer and gardener.

Some genial spirits have given the most elaborate attention to the question of the value of birds. One gentleman took his position some fifteen feet from the nest of an oriole, in the top of a peach-tree, to observe his habits. The nest contained four young ones, well fledged, which every now and then would stand upon the edge of the nest to try their wings. They were, therefore, at an age which required the largest supply of food. This the parents furnished at intervals of two to six minutes, throughout the day. They lighted on the trees, the vines, the grass, and other shrubbery, clinging at times to the most extreme and delicate points of the leaves, in search of insects. Nothing seemed to come amiss to these sharp eyed foragers – grasshoppers, caterpillars, worms, and the smaller flies. Sometimes one, and sometimes as many as six, were plainly fed to the young ones at once. They would also carry away the refuse litter from the nest, and drop it many yards off. A little figuring gives the result of this incessant warfare against the insects. For only eight working hours it will be 1000 worms destroyed by a single pair of birds. But if a hundred pairs be domesticated on the premises, the destruction will amount to 100,000 daily, or 3,000,000 a month !

This may seem to be a mere paper calculation, but the annals of ornithology are crowded with confirmatory facts. The robin is accused of appropriating the fruit which he has protected during the growing season from a cloud of enemies. But his principal food is spiders, beetles, caterpillars, worms, and larvæ. Nearly 200 larvæ have been taken from the gizzard of a single bird. He feeds voraciously on those of the destructive worm. In July he takes a few strawberries, cherries, and pulpy fruits generally, more as a dessert than any thing else, because it is invariably found to be largely intermixed with insects. Robins killed in the country, at a distance from gardens and fruit-trees, are found to contain less stone-fruit than those near villages; showing that this bird is not an extensive forager. If our choicest fruits are near at hand, he takes a small toll of them, but a small one only. In reality, a very considerable part of every crop of grain and fruit is planted, not for the mouths of our children, but for the fly, the curculio, and the canker-worm, or some other of these pests of husbandry. Science has done something, and will no doubt do more, to alleviate the plague. It has already taught us not to wage equal war on the wheat-fly and the parasite which preys upon it; and it will, perhaps, eventually persuade those who need the lesson, that a few peas and cherries are well bestowed by way of dessert on the cheerful little warblers, who turn our gardens into concert-rooms, and do so much to aid us in the warfare against the grubs and caterpillars, which form their principal meal.

But if the subject of the value of insect-destroying birds has been so much overlooked in this country, it is not so in Europe. It has been brought formally before the French Senate, and is now before the French government. Learned commissioners have reported upon it, and it is by no means improbable that special legislation will presently follow. The inquiry has been conducted with an elaborate accuracy characteristic of French legislation. Insects and birds have been carefully classified according to their several species; their habits of feeding have been closely observed, and the results ascertained and computed. It has been concluded that by no agency, save that of little birds, can the ravages of insects be kept down. There are some birds which live exclusively upon insects and grubs, and the quantity which they destroy is enormous. There are others which live partly on grubs, and partly on grain, doing some damage, but providing an abundant compensation. A third class – the Birds of Prey – are excepted from the category of benefactors, and are pronounced, too precipitately we think, to be noxious, inasmuch as they live mostly upon the smaller birds. One class is a match for the other. A certain insect was found to lay 2,000 eggs, but a single tom-tit was found to eat 200,000 eggs a year. A swallow devours about 543 insects a day, eggs and all. A sparrow's nest, in the city of Paris, was found to contain 700 pairs of the upper wings of cockchafers, though, of course, in such a place food of other kinds was procurable in abundance. It will easily be seen, therefore, what an excess of insect life is produced when a counterpoise like this is withdrawn; and the statistics before us show clearly to what an extent the balance of nature has been disturbed. A third, and wholly artificial class of destroyers has been introduced. Every *chasseur*, during the season, kills, it is said, from 100 to 200 birds daily. A single child has been known to come home at night with 100 birds' eggs, and it has been calculated and reported that the number of birds' eggs destroyed annually in France is between 80,000,000 and 100,000,000. The result is, that little birds in that country are actually dying out; some species have already disappeared, and others are rapidly diminishing. But there is another consequence. The French crops have suffered terribly from the superabundance of insect vermin. Not only the various kinds of grain, but the vines, the olives, and even forest trees, tell the same tale of mischief, till at length the alarm has become serious. Birds are now likely to be protected; indeed their rise in public estimation has been signally rapid. Some philosopher has declared, and the report quotes the saying as a profound one, that " the birds can live without man, but man cannot live without the birds."

The same results are being experienced in this country, and our whole agricultural press, as well as the experience of every fruit-grower and gardener, testifies to the fact that our fruit is disappearing as the birds upon our premises are permitted to perish. Every humane and prudent man will therefore do his utmost to preserve them.

CHAPTER XVI.
CLOSE OF MY FIRST YEAR –
ITS LOSS AND GAIN.

It was now the dead of winter. Every thing was frozen up; but though cheerless without, it was far from being so within. My little library, well supplied with books and the literature of the day, afforded me an intellectual banquet which never palled upon the appetite. Here my desk was ever open; here pen, and ink, and diary were constantly at hand, for entering down my expenditures and receipts, with facts and observations for future use. Thus conveniently provided, and all my life accustomed to accounts, I found no difficulty at the year's end in ascertaining to a dollar whether my first season's experience had been one of loss or gain. I give the particulars in full –

Cost of stable manure and ashes	$248.00
Plaster and guano, not all used	20.00
Ploughing, harrowing, and digging up the garden	30.00
Cabbage and tomato plants	30.00
Loss on my first cow	7.00
Garden seeds	8.00
Cost of six pigs	12.00
Com-meal and bran	28.00
Dick's wages for six months	72.00
	$455.00

Here was an outlay of $455, all of which was likely to occur every year, except the two items of loss on cow, and cost of buying cabbages and tomato-plants, which have subsequently been raised in a hotbed at home, without costing a dollar. The great item is in manure, amounting to $268; and this must be kept at the same figure, if not increased, unless an equal quantity can, by some process, be manufactured at home.

Then there was the following permanent outlay made in stocking the farm with fruit:

Strawberries for six acres $120.00
Raspberries for two acres 34.00
804 Peach-trees, and planting them 72.36

$226.36

This constituted a permanent investment of capital, and would not have to be repeated, so that the actual cost the first year was, as stated, $455. My own time and labor are not charged, because that item is adjusted in the grand result of whether the farm supported me or not. There was also the cost of horse and cow, ploughs, and other tools; but these, too, were investments, not expenses. They could be resold for money, no doubt, at some loss. A portion of that capital could therefore be recovered. So, also, with the large item of $226.36, invested in standard fruits; as, if the farm were sold, its being stocked with them would insure its bringing a higher price in consequence, probably enough to refund the capital thus invested.

It is fair, therefore, to charge the current expenses only against the current receipts. The latter were as follows:

Sales of blackberry plants $460.00
" cabbages 82.00
" tomatoes 120.00
" garden products 80.00
" pork 49.00

$791.00

Current expenses, as stated 455.00
Profit 336.00

This was about $1.25 per day for the two hundred and seventy-five days we had been in the country, from April 1st to January 1st, and, when added to our copious supplies of vegetables, fruit, pork, and milk, it kept the family in abundance. I proved -this by a very simple formula. I knew exactly how much cash I had on hand when I began in April, and from that amount deducted the cost of all my permanent investments in standard fruits, stock, and implements, and found that the remainder came within a few cents of the balance on hand in January. I did not owe a dollar, and had food enough to keep my stock till spring. The season had been a good one for me, and we felt the greatest encouragement to persevere, as the first difficulties had been overcome, and the second season promised to be much more profitable. I considered the problem as very nearly solved.

It will be noted that no cash was received for strawberries, and herein is involved a fact important to be known and acted on by the growers of this fruit. Most men, when planting them, say in March or April, are impatient for a crop in June. But this should never be allowed. As soon as the blossoms appear, they should be removed. The newly transplanted vine has work enough thrown upon its roots in repairing the damage it has suffered in being removed from one location to another, without being compelled, in addition, to mature a crop of fruit. To require it to do both is imposing on the roots a task they are many times unable to perform. The draft upon them by the ripening fruit is more than they can bear. I have known large fields of newly-planted vines perish in a dry season from this cause alone. The writers on strawberry culture sometimes recommend removing the blossoms the first year, but not with sufficient urgency. I lay it down as absolutely indispensable to the establishment of a robust growth. Thus believing, my blossoms were all clipped off with scissors; and hence, though stronger plants were thus produced, yet there was no fruit to sell.

It must also be remembered that my entire profit consisted of the single item of sales of plants; hence, if there had been no demand for Lawtons, or if I had happened to have none for sale, there would have been an actual loss. My having them was a mere accident, and my luck in this respect was quite exceptional. Unless others happen to be equally lucky, they may set down their first year as very certain to yield no profit. With persons as inexperienced as I was when beginning, no other result should be expected.

Winter is proverbially the farmer's holiday. But it was no idle time with me. I had too long been trained to habits of industry, to lounge about the house simply because no weeds could be found to kill. The careful man will find a world of fixing up to do for winter. As it came on slowly through a gorgeous Indian summer, I set myself to cleaning up the litter round the premises, and put the garden into the best condition for the coming season. The verbenas had gone from the borders; the petunias had withered on the little mound whereon their red and white had flashed so gayly in captivating contrast during the summer; the delicate cypress-vine had blackened at the touch of a single frosty night; the lady-slipper hung her flowery head; all the family of roses had faded; the morning-glory had withered; even the hardy honeysuckle had been frozen crisp. From the fruit-trees a cloud of leaves had fallen upon every garden-walk. Plants that needed housing were carefully potted, and taken under cover. The walks were cleared of leaves by transferring them to the barnyard. Bushes, trees, and vines were trimmed. Every remnant of decay was removed. The December sunshine fell upon a garden so trim and neat, that even in the bleakest day it was not unpleasant to wander through its alleys, and observe those wintry visitants, the snow-birds, gathering from the bushes their scanty store of favorite seeds. The asparagus was covered deeply with its favorite manure, and heavily salted. Tender roses were banked up with barnyard scrapings, and every delicate plant protected for its long season of hybernation.

Dick had his share of exemption from excessive labor. But I kept him tolerably busy for weeks in gathering up the cloud of leaves which fell throughout the neighborhood from roadsides lined with trees. No manure is so well worth saving in October and November as the falling leaves. They contain nearly three times as much nitrogen as ordinary barnyard manure; and every gardener who has strewn and covered them in his trenches late in the fall or in December, must have noticed the next season how black and moist the soil is that adheres to the thrifty young beets he pulls. No vegetable substance yields its woody fibre and becomes soluble quicker than leaves; and, from this very cause, they are soon dried up, scattered to the winds, and wasted, if not now gathered and trenched in, or composted, before the advent of severe winter.

My horse, and cow, and pigs, all slept in leaves. Their beds were warm and easy, and the saving of straw for litter was an item. As they were abundant, and very convenient, Dick carted to the barnyard an enormous quantity. Placing enough of them under cover, he littered all the stock with them until spring. The remainder was composted with the contents of the barnyard, and thus made a very important addition to my stock of manure. Thus the leaf-harvest is one of importance to the farmer, if he will but avail himself of it. A calm day or two spent in this business will enable him to get together a large pile of these fallen leaves; and if stowed in a dry place, he will experience the good effects of them in the improved condition of his stock, compared with those which are suffered to lie down, and perhaps he frozen down, in their own filth. The fertilizing material of leaves also adds essentially to the enriching qualities of the manureheap. Gardeners prize highly a compost made in part of decomposed leaves. The leaf-harvest is the last harvest of the year, and should be thoroughly attended to at the proper time.

The leisure of the season gave us greater opportunity for intercourse, both at home and abroad. The city was comparatively at our door, as accessible as ever – we were really mere suburbans. We ran down in an hour to be spectators of any unusual sight, and frequently attended the evening lectures of distinguished men. It was impossible for the world to sweep on, leaving us to stagnate. How different this winter seemed to me from any preceding one ! Formerly, this long season had been one of constant toiling; now, it was one of almost uninterrupted recreation. How different the path I travelled from that in which ambition hurries forward – too narrow for friendship, too crooked for love, too rugged for honesty, and too dark for science ! Thus, if we choose, we may sandwich in the poetry with the prose of life. Thus, many a dainty happiness and relishing enjoyment may come between the slices of every-day work, if we only so determine.

CHAPTER XVII.
MY SECOND YEAR – TRENCHING
THE GARDEN – STRAW BERRY PROFITS.

WINTER having passed away, the time for labor and the singing of birds again returned. Long before the land in Pennsylvania was fit to plough, the admirable soil of New Jersey had been turned over, and planted with early peas. One of its most valuable peculiarities is that of being at all times fit for ploughing, except when actually frozen hard. Even after heavy rains, when denser soils require a fortnight's drying before getting into condition for the plough, this is ready in a day or two. Its sandy character, instead of being a disadvantage, is one of its highest recommendations. It is thus two to three weeks earlier in yielding up its ripened products for market. Peas are the first things planted in the open fields. The traveller coming from the north, when passing by rail to Philadelphia through this genial region, has been frequently surprised at seeing the young pea-vines peeping up above a thin covering of snow, their long rows of delicate green stretching across extensive fields, and presenting a singular contrast with the fleecy covering around them. Naturally hardy, they survive the cold, and as the snow rapidly disappears they immediately renew their growth.

Having been much surprised by the profit yielded last year from the garden, I was determined to give it a better chance than ever, and to try the effect of thorough farming on a limited scale. I accordingly set Dick to covering it fully three inches deep with well-rotted stable-manure, of which I had purchased in the city my usual quantity, §200 worth, though hoping that I could so contrive it hereafter as not to be obliged to make so heavy a cash outlay for this material. I then procured him a spade fifteen inches long in the blade, and set him to trenching every inch of it not occupied by standard fruits. These had luckily been arranged in rows in borders by themselves, thus leaving large, open beds, in which the operation of trenching could be thoroughly practised. I estimated the open ground to be very nearly half an acre. I began by digging a trench from one end of the open space to the other, three feet wide and two deep, removing the earth to the further side of the open space. Then the bottom of the trench was dug up with the fifteen-inch spade, and then covered lightly with manure.

The adjoining ground was then thrown in, mixing the top soil as we went along, and also abundance of manure, until the trench was filled. As the earth thus used was all taken from the adjoining strip of three feet wide, of course, when the trench was full, another of corresponding size appeared beside it. With this the operation was repeated until all the garden had been thoroughly gone over. The earth which had been removed from the first trench, went into the last one. But I was careful not to place the top soil in a body at the bottom, but scattered it well through the whole of the filling. If rich, the roots of every plant would find some portion of it, let them travel where they might. On the whole job we bestowed a great amount of care, but it was such a job as would not require repeating for years, and would be permanently beneficial. I thus deposited $50 worth of manure, as a fund of nourishment on which my vegetables could for a long time draw with certainty of profit.

Now, a surface soil of a few inches only, will not answer for a good garden. The roots of succulent vegetables must extend into a deeper bed of fertility; and a greater depth of pulverization is required to absorb surplus rains, and to give off the accumulated moisture in dry weather. A shallow soil will become deluged by a single shower, because the hard subsoil will not allow it to pass downward; and again, in the heat and drought of midsummer, a thin stratum is made dry and parched in a week, while one of greater depth becomes scarcely affected. I might cite numerous instances, besides my own, where trenched gardens remained in the finest state of luxuriance during the most severe droughts, when others under ordinary management were nearly burnt up with the heat, growth having quite ceased, and leaves curled and withering for want of moisture.

The mode of trenching must vary with circumstances. In small, circumscribed pieces of ground, necessity requires it to be done by hand, as has been just described. In large spaces the subsoil plough may be used, but not to equal benefit. There are many reasons why the soils of gardens should be made better than for ordinary farm-crops. Most of the products of gardens are of a succulent nature, or will otherwise bear high feeding, such as garden roots in general, plants whose leaves furnish food, as salad, cabbages, &c., or those which produce large and succulent fruits, as cucumbers, melons, squashes, &c. As nearly all garden crops are the immediate food of man, while many farm-crops are only the coarser food of animals, greater care and skill may properly be applied in bringing the former forward to a high degree of perfection. The great amount of family supplies which may be obtained from a half-acre garden, provided the best soil is prepared for their growth, renders it a matter of equal importance and economy to give the soil the very best preparation.

It rarely happens that there is much selection to be made in soils as we find them in nature, for gardening purposes, unless particular attention is given to the subject in choosing a site for a new dwelling.

Generally, we have to take the land as we find it. Unless, therefore, we happen to find it just right, we should endeavor to improve it in the best manner. The principal means for making a perfect garden soil, are draining, trenching, and manuring. Now, let none be startled at the outset with the fear of cost, in thus preparing the soil. The entire expense of preparing half an acre would not, in general, amount to more than the amount saved in a single year in the purchase of food for family supplies, by the fine and abundant vegetables afforded. If the owner cannot possibly prepare his half or quarter acre of land properly, then let him occupy the ground with something else than garden crops, and take only a single square rod (if he cannot attend to more), and give this the most perfect preparation. A square rod of rich, luxuriant vegetables, will be found more valuable than eighty rods, or half an acre of scant, dwarfed, and stringy growth, which no one will wish to eat; while the extra cost and labor spent on the eighty rods in seeds, digging, and hoeing, would have been more than sufficient to prepare the smaller plot in the most complete manner. Let the determination be made, therefore, at the commencement, to take no more land than can be properly prepared, and in the most thorough manner.

The ten peach-trees in the garden were thoroughly manured by digging in around them all the coal ashes made during the winter, first sifting them well. No stable manure was added, as it promotes too rank and watery a growth in the peach, while ashes of any kind are what this fruit most delights in. Then the butts were examined for worms, but the last year's application of tar had kept off the fly, and the old ravages of the enemy were found to be nearly healed over by the growth of new bark. A fresh coating of tar was applied, and thus every thing was made safe.

As the season advanced, my wife and daughter took charge of the garden, as usual, and with high hopes of greater success than ever. They had had one year's experience, while now the ground was in far better condition. Moreover, they seemed to have forgotten all about the weeds, as in calculating their prospective profits they did not mention them even once. I was careful not to do so, though I had my own suspicions on the subject. When the planting had been done, and things went on growing finely as the season advanced, they were suddenly reminded of their ancient enemy. The trenching and manuring had done as much for the weeds as for the vegetables. Why should they not? In her innocency, Kate thought the weeds should all have been buried in the trenches, as if their seeds had been deposited exclusively on the surface. But they grew more rampantly than ever during the entire season, and to my mind they seemed to be in greater quantity. But the fact worked no discouragement to either wife or daughter. They waged against them the same resolute warfare, early, late, and in the noonday sun, until Kate, in spite of a capacious sunbonnet, became a nut-brown maid. Not a weed was permitted to flourish to maturity.

The careful culture of the garden this year gave them even a better reward than it had done the year before. The failures of the last season were all avoided. Several kinds of seeds were soaked before being planted, which prevented failure and secured a quicker growth.- In addition to this, they raised a greater variety of vegetables expressly for the store; and with some, such as radishes and beets, they were particularly lucky, and realized high prices for all they had to dispose of. Then the high manuring and extra care bestowed upon the asparagus were apparent in the quick and vigorous shooting up of thick and tender roots, far more than we could consume, and so superior to any others that were taken to the store, that they sold rapidly at city prices. Thus they began to make sales earlier in the season, while their crops were far more abundant. The trenching and manuring was evidently a paying investment. In addition to all this, the season proved to be a good one for fruit. The garden trees bore abundantly. My ten peach-trees had by this time been rejuvenated, and were loaded with fruit. When as large as hickory nuts, I began the operation of removing all the smallest, and of thinning out unsparingly wherever they were excessively crowded. After going over five trees, I brought a bucketful of the expurgated peaches to my wife for exhibition. She seemed panic-stricken at the sight – protested that we should have no peaches that season, if I went on at that rate – besought me to remember my peculiar weakness for pies – and pleaded so eloquently that the other trees should not be stripped, as to induce me, much against my judgment, to suspend my ravages. Thus five had been thinned and five left untouched.

At the moment, I regretted her interference, but as compliance with her wishes always brought to me its own gratification, if not in one way, then in some other, so it did in this instance. In the first place, the peaches on the five denuded trees grew prodigiously larger and finer than those on the other five. I gathered them carefully and sent them to the city, where they brought me $41 clear of expenses, while the fruit from the other trees, sent to market with similar care, netted only $17, and those used in the family from the same trees, estimated at the same rates, were worth $9, making, on those five, a difference of $15 in favor of thinning. Thus, the ten produced $58; but if all had been thinned, the product would have been $82.

This unexpected result satisfied my wife ever afterwards that it was quality, and not mere quantity, that the market wanted. Her own garden sales would have convinced her of this, had she observed them closely; but having overlooked results there, it required an illustration too striking to be gainsayed, and this the peach-trees furnished. All these figures appear in Kate's account-book. I had provided her with one expressly, for the garden operations, a nice gold pen, and every other possible convenience for making entries at the moment any transaction occurred. I had also taught her the simplest form for keeping her accounts, and caused her to keep a pass-book with the store, in which every consignment should be entered, so that her book and the storekeeper's should be a check on errors that might be found in either. She thus became extremely expert at her accounts, and as she took especial interest in the matter, could tell from memory, at the week's end, how many dollars' worth of produce she had sold. I found the amount running up quite hopefully as the season advanced, and when it had closed, she announced the total to be $63 without the peaches, or $121 by including them.

But she had paid some money for seeds; as an offset to which, no cash had been expended in digging, as Dick and myself had done it all.

So much for the garden this year. On my nine acres of ploughed land there was plenty of work to be done. Our old enemy, the weeds, did not seem to have diminished in number, notwithstanding our slaughter the previous year. They came up as thick and vigorous as ever, and required quite as much labor to master them, as the hoe was offener required among the rows of raspberries and strawberries. My dogged fellow, Dick, took this matter with perfect unconcern – said he knew it would be so, and that I would find the weeds could not be killed – but he might as well work among them as at any thing else. I ceased to argue with him on the subject, and as I had full faith in coming out right in the end, was content to silently bide my time.

This year I planted an acre with tomatoes, having raised abundance of fine plants in a hotbed, as well as egg-plants for the garden. I set them out in rows, three and one-half feet apart each way, and manured them well, twice as heavily as many of my neighbors did. This gave me 3,760 plants to the acre. The product was almost incredible, and amounted to 501 bushels, or about five quarts a hill, a far better yield than I had had the first year. From some hills as many as ten quarts each were gathered. I managed to get twenty baskets into New York market among the very first of the season, where they netted me $60. The next twenty netted $25, the next twenty only $15, as numerous competitors came in, and the next thirty cleared no more. After that the usual glut came on, and down went the price to twenty and even fifteen cents. But at twenty and twenty-five I continued to forward to Philadelphia, where they paid better than to let them rot on the ground. From 200 baskets at these low prices I netted $35. Then, in the height of the season, all picking was suspended, except for the pigs, who thus had any quantity they could consume. But the glut gradually subsided as tomatoes perished on the vines, and the price again rose in market to twenty-five cents, then to fifty, then to a dollar, and upwards. But my single acre afforded me but few- at the close of the season. I did not manage to realize $40 from the fag-end of the year, making a total net yield of $190.

Others near me, older hands at the business, did much better, but I thought this well enough. I would prefer raising tomatoes at 37 cents a bushel to potatoes at 75. The amount realized from an acre far exceeds that of potatoes. A smart man will gather from sixty to seventy bushels a day. The expense of cultivating, using plenty of manure, is about $60 per acre, and the gross yield may be safely calculated $250, leaving about $200 sure surplus. If it were not for the sudden and tremendous fall in prices to which tomatoes are subject soon after they come into market, growers might become rich in a few years.

The other acre was occupied with corn, roots, and cabbage, for winter feeding, with potatoes for family use. Turnips were sowed wherever room could be found for them, and no spot about the farm was permitted to remain idle. A hill of corn, a cabbage, a pumpkin-vine, or whatever else was suited to it, was planted. But of potatoes we did sell enough to amount to $24. On the acre occupied with blackberries, early cabbages were planted to the number of 4,000. Many of these, of course, were small and not marketable, though well manured and carefully attended. But all such were very acceptable in the barnyard and pig-pen. Of sound cabbages I sold 3,120, at an average of two and one-quarter cents, amounting to $70.20. I cannot tell how it was, but other persons close to me raised larger and better heads, and of course realized better prices. But I had no reason to complain.

The strawberries came first into market. I had labored to allow no runners to grow and take root except such as were necessary to fill up the line of each row. Most of the others had been clipped off as fast as they showed themselves. Thus the whole strength of the plant was concentrated into the fruit. In other words, I set out to raise fruit, not plants; and my rows were, therefore, composed of single stools, standing about four to six inches apart in the row. The ground between the rows was consequently clear for the passage of the horse-weeder, which kept it nice and clean throughout the season, while there was no sort of difficulty in getting between the stools with either the hand, or a small hoe, to keep out grass and weeds. The stools were consequently strong and healthy, and stood up higher from the ground than plants which grow in matted beds, thus measurably keeping clear of the sand and grit which heavy rains throw up on berries that lie very near the ground. The truth is, the ground for a foot all round each stool ought to have had a covering of cut straw, leaves, or something else for the fruit to rest upon, thus to keep them clean, as well as to preserve them from drought. But I did not so well understand the question at that time as I do now.

The fruit ripened beautifully, and grew to prodigious size, larger than most we had ever seen. The several pickings of the first week yielded 600 quart boxes of the choicest fruit, which I dispatched by railroad to an agent in New York, with whom I had previously made arrangements to receive them. The greatest care was used in preparing them for market When taken from the vines they were put directly into the small boxes, and these carried to the house, where, under a large shed adjoining the kitchen, my wife and daughters had made preparations to receive them. Here they were spread out on a large pine table, and all the larger berries separated from the smaller ones, each kind being put into boxes which were kept separate from the other. The show made by fruit thus assorted was truly magnificent, and to the pleasure my wife experienced in handling and arranging it, she was constantly testifying. Thus 600 quarts of the finest fruit we had ever beheld, were sent the first week to New York. It was, of course, nearly ten days ahead of the season in that region – there could be no New York grown berries in marke. At the week's end the agent remitted me $300 clear of freight and commission ! They had netted me half a dollar a quart. I confess to having been greatly astonished and delighted – it was certainly twice as much as we had expected. When I showed the agent's letter to my wife, she was quite amazed. Kate, who had heard a good deal of complaint about high prices, while we lived in the city, after reading the letter, laid it down, observing –

"I think it will not do to complain of high prices now !"

"No," replied my wife, "the tables are turned. Half a dollar a quart ! How much I pity those poor people."

And as she said this, I handed her a quart bowl of the luscious fruit, which I had been sugaring heavily while she was studying out the figures in the agent's letter, and I feel persuaded no lover of strawberries ever consumed them with a more smacking relish.

The agent spoke in his letter of the admirable manner in which our berries were forwarded – all alike, all uniformly prime large fruit – not merely big ones on top of the box as decoys, and as the prelude to finding none but little runts at bottom. This established for us a reputation; our boxes could be guaranteed to contain prime fruit all through. Hence the agent could sell any quantity we could send. Indeed, it was impossible to send him too much. Thus we continued to pick 'over our vines from three to four times weekly. As the ripening of the fruit went on, the sight was truly marvellous to look at. When the season was at its height, the ground seemed almost red with berries. Then the famous doctrine of squatter sovereignty was effectually carried out on my premises, for there were twenty girls and boys upon their knees or hams, engaged in picking berries at two cents a quart. Industrious little toilers they were, many of them earning from one to two dollars daily. Some pickers were women grown, some widows, some even aged women. It was a harvest to them also.

The small boxes were packed in chests each holding from twenty-four to sixty, just nicely filling the chest, so that there should be no rattling or shaking about, or spilling over of the fruit. The lid, when shut down and fastened, held all snug. These chests were taken to the railroad station close by, the same afternoon the berries were picked, and reached New York the same night. The agents knowing they were coming, had them all sold before they arrived, and immediately delivering them to the purchasers, they in turn delivered to their customers, and thus in less than twenty-four hours from the time of leaving my ground, they were in the hands of the consumers. This whole business of conveying fruit to distant markets by steamboat and rail, is thor oughly systematized. It is an immense item in the general freight-list of the great seaboard railroads, constantly growing, and as surely enriching both grower and carrier. For the former it insures a sale of all his products in the highest markets, and in fact brings them to his very door.

Before the building of the Camden and Amboy Railroad no such facilities existed, and consequently not a tenth of the fruit and truck now raised in New . Jersey was then produced. But an outlet being thus established, production commenced. Farms were manured, their yield increased, and stations for the receipt of freight were built at every few miles along the railroad. They continue to increase in number up to this day. Lands rose in value, better fences were supplied, new houses built, and the whole system of county roads was revolutionized. As every thing that could be raised now found a cash market, so every convenience for getting it there was attended to. Hence, gravel turnpikes Avere built, which, stretching back into the country, enabled growers at all seasons to transport their products over smooth roads to the nearest station. These numerous feeders to the great railroad caused the income from way-traffic to increase enormously. All interests Avere signally benefited, and a new career of improvement for NEW Jersey was inaugurated. The farmers became rich on lands which for generations had kept their former owners poor.

My agents Avere punctual in advising me by the first mail, and sometimes by telegraph, of the sale and price of each consignment, thus keeping me constantly posted up as to the condition of the market. They paid the freight on each consignment, deducted it from the proceeds, and returned the chests, though sometimes with a few small boxes missing, a loss to which growers seem to be regularly subjected, so long as they use a box which they eannot afford to give away with the fruit. I thus fed the northern cities as long as the price was maintained. But, as is the case with all market produce, prices gradually declined as other growers came in, for all hands sought to sell in the best market. As the end of the season is generally a period of very low prices, it must be counteracted by every effort to secure high ones at the beginning, in this way maintaining a remunerative average during the whole. Thus, the half dollar per quart which I obtained for the first and best, by equalization with lower prices through the remainder of the season, was unable to raise the average of the whole crop above sixteen cents net. But this abundantly satisfied me, as I sent to market 5,360 quarts, thus producing $857.60. Besides these, we had the satisfaction of making generous presents to some particular friends in the city, while at home we rioted upon them daily, and laid by an extraordinary quantity in the shape of preserves for winter use, a luxury which we had never indulged in during our residence in the city. I may add that during the whole strawberry season it was observed that our city friends seemed to take an extraordinary interest in our proceedings and success. They came up to see us even more numerously than during the dog-days, and no great effort was required, no second invitation necessary, to induce them to prolong their visits. But we con sidered them entirely excusable, as the strawberries and cream were not only unexceptionable, but abundant. However, I must confess, that in the busiest part of the season our female visitors rolled up their sleeves, and fell to with my wife and daughters for hours at a time, aiding them in assorting and boxing the huge quantities of noble fruit as it came in from the field.

In order to send this fruit to market, I was obliged to purchase 3,000 quart boxes, and 50 chests to contain them. These cost me $200.

I could not fill all the boxes at each picking, but as one set of boxes was away off in market, it was necessary for me to have duplicates on hand, in which to pick other berries as they ripened, without being compelled to wait until the first lot of boxes came back. Sometimes it was a week or ten days before they were returned to me, according as the agent was prompt or dilatory. Thus, one supply of boxes filled with fruit was constantly going forward, while another of empty ones was on the way back. So extensive has this berry business become, that I could name parties who have as much as $500 to $1,500 invested in chests and boxes for the transportation of fruit to market. But their profits are in proportion to the extent of their investment.

While on this subject of boxes for the transportation of fruit to distant markets, a suggestion occurs to me which some ingenious man may be able to work up into profitable use. It is sometimes quite a trouble for the grower to get his chests returned at the proper time. Sometimes the agent is careless and inattentive, keeps them twice as many days as he ought to, when the owner really needs them. Sometimes an accident on the railroad delays their return for a week or ten days. In either case, the grower is subjected to great inconvenience; and if his chests fail to return at all, his ripened fruit will perish on his hands for want of boxes in which to send them off. It is to be always safe from these contingencies that he finds it necessary to keep so large a quantity on hand. Then, many of the boxes are never returned, the chests coming back only half or quarter filled. All this is very unjustly made the grower's loss.

But a remedy for this evil can and ought to be provided. The trade needs for its use a box so cheap that it can afford to give it away. Then, being packed in rough, open crates, cheaply put together of common lath, with latticed sides, neither crates nor boxes need be returned. The grower will save the return-freight, and be in no danger of ever being short of boxes by the negligence of others. This is really a very urgent want of the trade. The agent sells by wholesale to the retailer, who takes the chest to his stand or store, where he sells the contents, one or more boxes to each customer. These sometimes have no baskets with them in which to empty the berries, and so the retailer, to insure a sale, permits the buyer to carry off the boxes, and the latter neglects to return them. In the same way they are sent to hotels and boarding-houses, where they are lost by hundreds. Again, the obligation imposed on a buyer to return the boxes to a retailer, is constantly preventing hundreds of chance purchasers of rare fruit from taking it; but if the seller could say to him that the box goes with the fruit, and need not be returned, the mere convenience of the thing would be sufficient to determine the sale of large quantities, – the purchaser would carry it home in his hand.

The maker of a cheap box like this would find the sale almost indefinite. It would be constant, and annually increasing. The same buyers would require fresh supplies every season. A mere chip box, rounded out of a single shaving, and just stiff enough to prevent the sides from collapsing, would answer every purpose. The pill-boxes which are made from shavings may serve as the model. Here is a great and growing want, which our countrymen are abundantly able to supply, and to which some of them cannot too soon direct their attention. If the cost of transmitting the boxes to the buyers be too great for so cheap a contrivance, then let the shavings be manufactured of the exact size required, and delivered in a flat state to the buyer, with the circular bottom, by him to be put together during the leisure days of winter. A single touch of glue will hold the shaving in position, and a couple of tacks will keep the bottom in its place. The whole affair being for temporary use, need be nothing more than temporary itself. A portion of the labor of manufacturing being done by the grower, will reduce the cost. If constructed as suggested, such boxes would be quite as neat as the majority now in use, while they would possess the charm of always being clean and sweet. Our country is at this moment full of machinery exactly fitted to produce them, much of it located in regions where timber and power are obtainable at the minimum cost. The suggestion should be appropriated by its owners at the earliest possible moment.

CHAPTER XVIII.
RASPBERRIES – THE LAWTONS.

To strawberries succeeded raspberries. My stock of boxes was thus useful a second time. But raspberries are not always reliable for a full crop the first season after planting, and so it turned out with mine. They bore only moderately; but by exercising the same care in rejecting all inferior specimens, the first commanded twenty-five cents a quart in market; gradually declining to twelve, below which none were sold. I marketed only 242 quarts from the whole, netting an average of 16 cents a quart, or $38.72. In price they were thus equal to strawberries. In addition to this, we consumed in the family as much as all desired, and that was not small. I had heard of others doing considerably better than this, but had no disposition to be dissatisfied.

The trade in raspberries is increasing rapidly in the neighborhood of all our large cities, stimulated by the establishment of steamboats and railroads, on which they go so quickly and cheaply to market. It is probably greater in New York State than elsewhere. The citizens of Marlborough, in Ulster county, have a steamboat regularly employed for almost the sole business of transporting their raspberries to New York. In a single season their sales of this fruit amount to nearly $90,000. The demand is inexhaustible, and the cultivation consequently increases. In the immediate vicinity of Milton, in the same county, there are over 100 acres of them, and new plantations are being annually established. The pickers are on the ground as soon as the dew is off, as the berries do not keep so well when gathered wet. I have there seen fifty pickers at work at the same time, men, women, and children, some of them astonishingly expert, earning as much as $2 in a day. Several persons were constantly employed in packing the neat little baskets into crates, the baskets holding nearly a pint. By six o'clock the crates were put on board the steamboat, and by sunrise next morning they were in Washington market. As many as 80,000 baskets are carried at a single trip. The retail price averages ten cents a basket, one boat thus carrying $800 worth in a single day. All this cultivation being conducted in a large way, the yield per acre is consequently less than from small patches thoroughly attended to. There are repeated instances of $400 and even $600 being made clear from a single acre of raspberries.

The culture in Ulster county, though at first view appearing small, yet gives employment to, and distributes its gains among thousands of persons. The mere culture requires the services of a large number of people. The pickers there, as well as in New Jersey, constitute a small army, there being five or more required for each acre, and the moneys thus earned by these industrious people go far towards making entire families comfortable during some months of the year. The season for raspberries continues about six weeks. Many of the baskets which are used about New York are imported from France. Frequently the supply is unequal to the demand. If the chip boxes were introduced, as suggested in the last chapter, the whole of this outlay to foreign countries could be stopped. It is strange, indeed, that any portion of our people should be compelled to depend on France for baskets in which to convey their berries to market.

As my raspberries disappeared, so in regular succession came the Lawton blackberries. I had cut off the tip of every cane the preceding July. This, by stopping the upward growth, drove the whole energy of the plant into the formation of branches. These had in turn been shortened to a foot in length at the close of last season. This process, by limiting the quantity of fruit to be produced, increased the size of the berries. I am certain of this fact, by long experience with this plant. It also prevented the ends of the branches resting on the ground, when all fruit there produced would otherwise be ruined by being covered with dust or mud. Besides, this was their first bearing year, and as they had not had time to acquire a full supply of roots, it would be unwise to let them overbear themselves. Some few which had grown to a great height were staked up with pickets four and a half feet long, and tied, the pickets costing $11 per thousand at the lumber-yard. But the majority did not need this staking up the first season; but many of the canes sent up this year, for bearers the next, it was necessary to support with stakes.

The crop was excellent in quality, but not large. I began picking July 20, and thus had the third use of my stock of boxes. I practised the same care in assorting these berries for market which had been observed with the others, keeping the larger ones separate from the smaller ones. Thus a chest of the selected berries, when exposed to view, presented a truly magnificent sight. Up to this time they had never been seen by fifty frequenters of the Philadelphia markets. But when this rare display was first opened in two of the principal markets, it produced a great sensation. None had been picked until perfectly ripe, hence the rare and melting flavor peculiar to the Lawton pervaded every berry. They sold rapidly and netted me thirty cents a quart, the smaller ones twenty-five cents. There appeared to be no limit to the demand at these prices. Buyers cheerfully gave them, though they could get the common wild blackberry in the same market at ten cents. Now, it cost me no more to raise the Lawtons than it would have done to raise the common article. But this is merely another illustration of the folly of raising the poorest fruit to sell at the lowest prices, instead of the best to sell at the highest.

The crop of Lawtons amounted to five hundred and ninety-two quarts, and netted me $159.84, an average of twenty-seven cents a quart. My family did not fail to eat even more than a usual allowance. As soon as the picking was done, while the plants were yet covered with leaves, Dick cut off at the ground all the canes which had just fruited, using a strong pair of snip-shears, which cut them through without any labor. These canes having done their duty would die in the autumn, could now be more easily cut than when grown hard after death, and if removed at once, would be out of the way of the new canes of this year's growth.

The latter could then be trimmed and staked up for the coming year, the removal of all which superfluous foliage would let in the sun and air more freely to the cabbages between the rows. The old wood being thus cut out, was gathered in a heap, and when dry enough was burned, the ashes being collected and scattered around the peach-trees. After this the limbs were all shortened in to a foot. They were very strong and vigorous, as in July the tops of the canes had all been taken off, leaving no cane more than four feet high. The branches were consequently very strong, giving promise of a fine crop another season. After this, such as needed it were staked up and tied, as the autumn and winter winds so blow and twist them about that otherwise they would be broken off. But subsequent practice has induced me to cut down to only three feet high; and this being done in July, when the plant is in full growth, the cane becomes so stiff and stocky before losing its leaves as to require no staking, and will support itself under any ordinary storm. I have seen growers of this fruit who neglected for two or three years, either from laziness or carelessness, to remove the old wood; but it made terrible work for the pickers, as in order to get at one year's fruit they were compelled to contend with three years' briers. Only a sloven will thus fail to re-move the old wood annually. I prefer removing it in the autumn, as soon as picking is over, for reasons above given, and also because at that time there is less to do than in the spring.

In the mean time the fame of the Lawton blackberry had greatly extended and the demand increased, but the propagation had also been stimulated. A class of growers had omitted tilling their grounds, so as to promote the growth of suckers, caring more for the sale of plants than for that of fruit. Hence the quantity to meet the demand was so large as to reduce the price, but I sold of this year's growth enough plants to produce me $213.50. Of this I laid out $54 in marl, which I devoted exclusively to the blackberries. I had been advised by a friend that marl was the specific manure for this plant, as of his own knowledge he knew it to be so. A half-peck was spread round each hill, and the remainder scattered over the ground. A single row was left unmarled. It showed the power of this fertilizer the next sea-son, as the rows thus manured were surprisingly better filled with fruit than that which received none. Since that I have continued to use this fertilizer on my blackberries, and can from experience recommend its use to all who may cultivate them.

With the sale of pork, amounting to $58, the receipts of my second year terminated. My cashbook showed the following as the total of receipts and expenditures:

Paid for stable manure	$200 00
Ashes, and Baugh's rawbone superphosphate	92.00
Marl	54.00
Dick's wages	144.00
Occasionai help	94.00
Feed for stock	79.30
Pigs bought	12.00
Garden and other seeds	13.00
Lumber, nails, and sundries	14.50
Stakes and twine	7.00
	$709.80

The credit side of the account was much better than last year, and was as follows:

From strawberries, 6 acres	$857.60
" Lawton blackberries, 1 acre	159.84
" Lawton plants	213.50
" raspberries, 2 acres	38.72
" tomatoes, 1 acre	190.00
" cabbages	70.20
" garden	63.00
" peaches, 10 trees in garden	58.00
" potatoes	24.00
" pork	58.00
" calf	2.00
	$1,734.86

The reader will not fail to bear in mind that in addition to this cash receipt towards the support of a family, we had not laid out a dollar for fruits or vegetables during the entire year. Having all OI them in unstinted abundance, with a most noble cow, the cash outlay for the family was necessarily very small; for no one knows, until he has all these things without paying for them in money, how very far they go towards making up the sum total of the cost of keeping a family of ten persons. In addition to this, we had a full six months' supply of pork on hand.

The reader will also be struck with the enormous difference in favor of the second year. But on dissecting the two accounts he will see good reason for this difference. In the first place, some improvement was natural, as the result of my increase of knowledge, – I was expected to be all the time growing wiser in my new calling. In the second place, some expenses incident to the initiatory year were lopped off; and third, three of my standard fruits had come into bearing. The increase of receipts was apparently sudden, but it was exactly what was to be expected. I used manure more freely, and on my acre of clover was particular to spread a good dressing of solid or liquid manure immediately after each mowing, so as to thus restore to it a full equivalent for the food taken away. This dressing was sometimes ashes, sometimes plaster, or bone-phosphate, or liquid, and in the fall a good topping from the barnyard. In return for this, the yield of clover was probably four times what it would have been had the lot been pastured and left unmanured. In fact, it became evident to me that the more manure I was able to apply on any crop, the more satisfactory were my returns. Hence, the soiling system was persevered in, and we had now become so accustomed to it that we considered it as no extra trouble.

The result of this year's operations was apparently conclusive. My expenses for the farm had been $709.80, while my receipts had been $1,734.86, leaving a surplus of $1,025.06 for the support of my family. But more than half of their support had been drawn from the products of the farm; and, at the year's end, when every account had been settled up, and every bill at the stores paid off, I found that of this $1,025.06 I had $567 in cash on hand, – proving that it had required only $458.06 in money, in addition to what we consumed from the farm, to keep us all with far more comfort than we had ever known in the city. Thus, after setting aside $356.06 for the purchase of manure, there was a clear surplus of $200 for investment.

I had never done better than this in the city. There, the year's end never found me with accounts squared up, and a clear cash balance on hand. Few occupations can be carried on in the city after so snug a fashion. Credit is there the rule, and cash the exception, – at least it was ten years ago. But in the apparently humbler trade of trucking and fruit-growing every thing is cash. Manure, the great staple article to be bought, can be had on credit; but all you grow from it is cash. Food must be paid for on delivery, and he who produces it will have no bad debts at the year's end but such as may exist from his own carelessness or neglect. Thus, what a farmer earns he gets. He loses none of his gains, if he attends to his business. They may be smaller, on paper, than those realized by dashing operators in the city, but they are infinitely more tangible; and if, as in my case, they should prove to be enough, what matters it as to the amount? The producers of food, therefore, possess this preponderating advantage over all other classes of business men: they go into a market where cash without limit is always ready to be paid down for whatever they bring to it. A business which is notoriously profitable, thus kept up at the cash level, and consequently free from the hazard of bad debts, cannot fail to enrich those who pursue it extensively, and with proper intelligence and industry. I could name various men who, beginning on less than a hundred dollars, and on rented land, have in a few years become its owners, and in the end arrived at great wealth, solely from the business of raising fruit and truck.

CHAPTER XIX.
LIQUID MANURES – AN ILLUSTRATION.

No sooner had the autumn of my second year fairly set in, and the leaves fallen, than I turned my attention more closely than ever to the subject of providing an abundant supply of manure, in hopes of being able to devise some plan by which to lessen the large cash outlay necessary to be annually made for it. I did not grudge the money for manure, any more than the sugar on my strawberries. Both were absolutely necessary; but economy in providing manure was as legitimate a method of increasing my profits as that of purchasing it. I knew it must be had in abundance: the point was, to increase the quantity while diminishing the outlay. Thus resolved, I kept Dick more actively at work than ever in gathering leaves all over the neighborhood, and when he had cleaned up the public roads, I then sent him into every piece of woods to which the owner would grant me access. In these he gathered the mould and half-rotted leaves which thickly covered the ground. I knew that he would thus bring home a quantity of pestiferous seeds, to plague us in the shape of weeds, but by this time we had learned to have no fear of them. By steadily pursuing this plan when no snow lay on the ground, he piled up in the barnyard a most astonishing quantity of leaves. There happened to be but little competition in the search for them, so that he had the ground clear for himself. All this addition to the manure heap cost me nothing. To this I added many hogsheads of bones, which the small boys of the neighborhood gathered up from pig-pens, slaughter-houses, and other places, and considered themselves well paid at ten cents a bushel for their labor. These were laid aside until the best and cheapest method could be devised for reducing them to powder, and so fitting them for use.

In the mean time, I frequently walked for miles away into the country, making acquaintance with the farmers, observing their different modes of cultivation, what crops they produced, and especially their methods of obtaining manures. As before observed, farmers have no secrets. Hence many valuable hints were obtained and treasured up, from which I have subsequently derived the greatest advantage. Some of these farmers were living on land which they had skinned into the most squalid poverty, and were on the high-road to being turned off by the sheriff. Others were manured at a money cost which astonished me, exceeding any outlay that I had made, but confirming to the letter all my preconceived opinions on the subject, that one acre thoroughly manured is worth ten that are starved. Of one farmer I learned particulars as to the history of his neighbor, which I felt a delicacy in asking of the latter himself. Some instances of success from the humblest beginnings were truly remarkable; but in all these I found that faith in manure lay at the bottom.

One case is too striking to be omitted. A German, with his wife, and two children just large enough to pull weeds and drive a cow, had settled, seven years before, on eight acres, from which the owner had been driven by running deeply in debt at the grog-shop. The drunkard's acres had of course become starved and desolate; the fences were half down, there was no garden, and the hovel, in which his unhappy family was once snugly housed, appeared ready to take its departure on the wings of the wind. Every fruit-tree had died. In this squalid condition the newly arrived German took possession, with the privilege of purchasing for $600. His whole capital was three dollars. He began with four pigs, which he paid for in work. The manure from these was daily emptied into an empty butter-firkin, which also served as a family water-closet, and the whole was converted into liquid manure, which was supplied to cabbages and onions. A gentleman who lived near, and who noted the progress of this industrious man, assured me that even in the exhausted soil where the crops were planted, the growth was almost incredible. On turnips and ruta-bagas the effect was equally great. Long before winter set in, this hero had bought a cow, for while his own crops were growing he had earned money by working around the neighbor-

hood. He readily obtained credit at the store, for he was soon discovered to be deserving. When away at work, his wife plied the hoe,
and acted as mistress of the aforesaid butter-tub, while the children
pulled weeds. His cabbages and roots exceeded any in the township;
they discharged his little store-bills, and kept his cow during the winter, while the living cow and the dead pigs kept the entire family, for
they lived about as close to the wind as possible.

This man's passion was for liquid manure. If he had done so
much with a tub, he was of course comparatively rich with a cow.
Then he sunk a hogshead in the ground, conducted the wash of the
kitchen into it, and there also emptied the droppings from the cow.
It was water-closet for her as well as for the family. It is true that few
of us would fancy such a smelling-bottle at the kitchen door; but it
never became a nuisance, for he kept it innoxious by frequent applications of plaster, which improved as well as purified the whole
contents. It was laborious to transport the fluid to his crops, but a
wheelbarrow came the second year to lessen the labor. There happened, by the merest accident, to be a quarter of an acre of raspberries surviving on the place. He dug all round these to the depth
of eighteen inches, trimmed them up, kept out the weeds, and gave
them enormous quantities of liquid manure. The yield was most extraordinary, for the second year of his location there he sold $84
worth of fruit. This encouraged him to plant more, until at the end of
four years he had made enough, from his raspberries alone, not only
to pay for his eight acres, but to accumulate a multitude of comforts
around him. In all this application of liquid manure his wife had aided him with unflagging industry.

It was natural for me to feel great interest in a case like this, so I
called repeatedly to see the grounds and converse with the German
owner. As it was seven years from his beginning when I first became
acquainted with him, his little farm bore no resemblance to its condition when he took possession. There were signs of thrift all over it. His
fences were new, and clear of hedge-rows; his house had been completely renovated; he had built a large barn and cattle-sheds, while

his garden was immeasurably better than mine. Every thing was in a condition exceeding all that I had seen elsewhere. His two girls had grown up into handsome young women, and had been for years at school. All this time he had continued to enlarge his means of manufacturing and applying liquid manure, as upon its use he placed his main dependence. He had sunk a large brick cistern in the barnyard, into which all the liquor from six cows and two horses was conducted, as well as the wash from the pig-pen and the barnyard. A fine pump in the cistern enabled him to keep his manure heap constantly saturated, the heap being always under cover, and to fill a hogshead mounted on wheels, from which he discharged the contents over his ground. The tub and underground hogshead with which he commenced were of course obsolete. If it be possible to build a monument out of liquid manure, here was one on this farm of eight acres. Its owner developed another peculiarity – he had no desire to buy more land.

This man's great success in a small way could not have been achieved without the most assiduous husbanding of manure, and this husbanding was accomplished by soiling his cow. As he increased his herd he continued the soiling system; but as it required more help, so he abandoned working for others and hired whatever help was necessary. The increase of his manure heap was so great that his little farm was soon brought into the highest possible condition. In favorable seasons he could grow huge crops of whatever he planted. But his progress was no greater than has repeatedly been made by others, who thoroughly prosecute the soiling system.

A frequent study of this remarkable instance of successful industry, led me to conclude that high farming must consist in the abundant use of manure in a liquid state. A fresh reading of forgotten pages shed abundance of new light upon the subject. The fluid excretia of every animal is worth more than the solid portion; but some are not contented with losing the fluid portions voided by the animals themselves, but they suffer the solid portions of their manure to undergo destructive fermentation in their barnyards, and thus to become soluble, and part, by washing, with the more valuable portions. Now it is well known that the inorganic matter in barnyard manure is always of a superior character, therefore valuable as well as soluble; and this is regularly parted with from the soil by those who permit the washings to be wasted by running off to other fields or to the roadside. I have seen whole townships where every barnyard on the roadside may be found discharging a broad stream of this life-blood of the farm into the public highway. The manure heap must be liquefied before the roots of plants can be benefited by the food it contains. No portion of a straw decomposed in the soil can feed a new plant until it is capable of being dissolved in water; and this solution cannot occur without chemical changes, whose conditions are supplied by the surroundings. Such changes can be made to occur in the barnyard by saturating the compost heap with barnyard liquor. All that nature's laws would in ten years effect in manures in an ordinary state, when ploughed into the ground, are ready, and occur in a single season, when the manures are presented to the roots of plants in a liquid form.

A suggestion appropriate to this matter may be made for the consideration of ingenious minds. Every farmer knows that a manure heap, when first composted, abounds in clods of matted ingredients so compact, that time alone will thoroughly reduce them to that state of pulverization in which manure becomes an available stimulant to the roots of plants. Fermentation, the result of composting or turning over a manure heap, does measurably destroy their cohesion, but not sufficiently. Few can afford to let their compost heaps remain long enough for the process of pulverization to become as perfect as it should be. Hence it is taken to the field still composed of hard clods, around which the roots may instinctively cluster, but into which they vainly seek to penetrate. Some careful farmers endeavor to remedy this defect by laboriously spading down the heap as it is carted away. The operation is a slow one, and does not half prepare the manure for distribution. A year or two is thus required for these clods to become properly pulverized, for they remain in the soil inert and. useless until subsequent ploughings and harrowing reduce them to powder.

As farmers cannot wait for time to perform this office in the manure heap, they should have machinery to do the work. A wooden cylinder, armed with long iron teeth, and revolving rapidly in a horizontal position, with the manure fed in at the top through a capacious hopper, would tear up the clods into tatters, and deliver the whole in the exact condition of fine powder, which the roots of all plants require. To do this would require less time and labor than the present custom of cutting down with either spade or drag. Better still, if the manure could be so broken up as it is taken from the barnyard to the compost heap; the process of disintegration thus begun would go on through the entire mass, until, when carted away, it would be found almost as friable as an ash heap. It is by contact of the countless mouths of the roots with minute particles of manure that they suck up nutriment, not by contact with a dense clod. Hence the astonishing and immediate efficacy of liquid manure. In that the nutriment has been reduced to its utmost condition of divisibility, and when the liquid is applied to the soil, saturation reaches the entire root, embracing its marvellous network of minute fibres, and affording to each the food which it may be seeking.

We cannot use liquid manures on a large scale, but thorough pulverization of that which is solid is a very near approach to the former. Immerse a compact clod in water, and the latter will require time to become discolored. But plunge an equal bulk of finely pulverized manure into water, and discoloration almost instantly occurs. Diffusion is inevitable from contact with the water. Now as rain is water, so a heavy shower falling on ground beneath which great clods of manure have been buried, produces in them no more liquefaction than it does on that which has been dropped in a bucket. On the other hand, if the ground be charged with finely pulverized manure, a soaking rain will immediately penetrate all its comminuted particles, extract the nutriment, and deliver it, properly diluted, into the open mouths of the millions of little rootlets which are waiting for it. Practically, this is liquid manure on the grandest scale. But no one can quickly realize its superior benefits from a newly buried compost heap, unless the latter has been effectually pulverized before being deposited either in or upon the ground.

I was so impressed by the example of the thriving German referred to, that I resolved to imitate him. He had given me a rich lesson in the art of manufacturing manures cheaply, though I thought it did not go far enough. Yet I made an immediate beginning by building a tank in the barnyard, into which the wash from stable, pig-pen, and yard was conducted. This was pumped up and distributed over the top of the manure heap under the shed, once or twice weekly. A huge compost heap was made of leaves, each layer being saturated with the liquor as the heap accumulated, so that the whole mass was moist with fluid manure. It was never suffered to become dry. Now, as in the centre of a manure heap there is no winter, decomposition went on at a rapid rate, especially among the leaves, stimulated by the peculiar solvents contained in the liquor. Thus, when taken out for use in the spring, both heaps had become reduced to a half fluid mass of highly concentrated manure, in a condition to be converted, under the first heavy rain, into immediate food for plants. Though my money-cost for manure for next season would be greater than before, yet my home manufacture was immense. As I was sure that high manuring was the key to heavy crops and high profits, so my studies, this winter, were as diligently pursued in the barnyard as in the library, and I flattered myself that I had gathered hints enough among my neighbors to enable me, after next year, to dispense entirely with the purchasing of manure.

But I had other reasons for avoiding the purchase of manure – none can be purchased clear of seeds, such as grass and weeds. I have already suffered severely from the foul trash that has been sold to me. One strong warning of the magnitude of the nuisance was given by the condition of my strawberries. A small portion of them was covered, at the approach of winter, with litter from the barnyard, and another portion with cornstalks. The object was protection from the cold; and it may be added that the result, so far as protection goes, was very gratifying. But when the covering was removed in April, the ground protected by the barnyard litter was found to be seeded with grass and other seeds, while that protected by the cornstalks was entirely clean. During a whole year I had the utmost difficulty to get the first piece of ground clear of these newly planted pests, and am sure that the labor thus exerted cost more than the strawberries were worth. From this sore experience I have- learned never to cover this fruit with barn-yard litter. When they are covered, cornstalks alone are used. They are drawn back into the balks in April, where they serve as a mulch to keep down the weeds, and ultimately decay into manure. Though not so neat to look at, nor so convenient to handle as straw, yet they answer quite as well, and at the same time cost a great deal less.

CHAPTER XX.
MY THIRD YEAR – LIQUID MANURE – THREE YEARS' RESULTS.

As usual with me at the opening of spring, the garden received our first attention. Dick covered it heavily with manure, cleared it up and made all ready for wife and daughter. This year we had no seeds to purchase, having carefully laid them aside from the last. In order to try for myself the value of liquid manuring, I mounted a barrel on a wheelbarrow, so that it could be turned in any direction, and the liquor be discharged through a sprinkler with the greatest convenience. Dick attended faithfully to this department. As early as January he had begun to sprinkle the asparagus; indeed he deluged it, putting on not less than twenty barrels of liquor before it was forked up. It had received its full share of rich manure in the autumn: the result of both applications being a more luxuriant growth of this delightful vegetable than perhaps even the Philadelphia market had ever exhibited. The shoots came up more numerously than before, were whiter, thicker, and tenderer, and commanded five cents a bunch more than any other. As the bed was a large one, and the yield great, we sold to the amount of $21. I certainly never tasted so luscious and tender an article. Its superiority was justly traceable, to some extent, to the liquid manure.

The same stimulant was freely administered all over the garden, and with marked results. It was never used in dry weather, nor when a hot sun was shining. We contrived to get it on at the beginning of a rain, or during drizzly weather, so that it should be immediately diluted and then carried down to the roots. I have no doubt it promoted the growth of weeds, as there was certainly more of them to kill this season than ever before. But we had all become reconciled to the sight of weeds – expected them as a matter of course – and my wife and Kate became thorough converts to Dick's heresy as to the impossibility of ever getting rid of them. I was pained to hear of this declension from what I regarded as the only true faith; but when I saw the terrible armies which came up in the garden just as regularly as Dick distributed his liquor, I confess they had abundant reason for the faith that was in them.

But the barnyard fluid was a good thing, notwithstanding. It brought the early beets into market ten days ahead of all competitors, thus securing the best prices. It was the same with radishes and salad. The latter is scarcely ever to be had in small country towns, and then only at high rates. But whether it was owing to the liquor or not, I will not say, but it came early into market in the best possible condition; and as there happened to be plenty of it, we sold to the amount of $19 of the very early, and then, as prices lowered, continued to send it to the store as long as it commanded two cents a head, after which the cow and pigs became exclusive customers. The fall vegetables, such as white onions, carrots and parsnips, having had more of the liquor, did even better, for they grew to very large size. It was the same thing with currants and gooseberries. The whole together produced $83; to which must be added the ten peach-trees, all which I had thinned out when the fruit was the size of hickory nuts, and with the same success as the previous year. This was in 1857, that time of panic, suspension, and insolvency. That year had been noted, even from its opening, as one of great scarcity of money in the cities, when all unlucky enough to need it were compelled to pay the highest rates for its use. But we in the country, being out of the ring, gave way to no panic, felt no scarcity, experienced no insolvency. Peaches brought as high a price as ever; as, let times in the city be black as they may, there is always money enough in somebody's hands to exchange for all the choice fruit that goes to market. The fruit from the ten trees produced me $69, making the whole product of the garden $152. I thought this was not doing well enough, and resolved to do better another year.

At the usual season for the weeds to show themselves on the nine acres, it very soon became evident that two years' warfare had resulted in a comparative conquest. It may be safely said that there was not half the usual number, and so it continued throughout the season. But no exertion was spared to keep them under, none being allowed to go to seed. This watchfulness being continued from that day to this, the mastery has. been complete. We still have weeds, but are no longer troubled with them as at the beginning. The secret lies in a nutshell – let none go to seed. Nor let any cultivator be discouraged, no matter how formidable the host he may have to attack at the beginning. But if he will procure the proper labor-saving tools, and drive them with a determined perseverance, success is sure.

As usual, the strawberries came first into market, and were prepared and sent off with even more care than formerly. The money pressure in the cities caused no reduction in price, and my net receipts were $903. An experienced grower near me, with only four acres, cleared $1,200 the same season. His crop was much heavier than mine. If he had practised the same care in assorting his fruit for market, he would have realized several hundred dollars more. But his effort was for quantity, not quality.

A portion of the raspberries had been thoroughly watered with the liquid manure, all through the colder spring months. It was too great a labor, with a single wheelbarrow, to supply the whole two acres, or it would have been similarly treated. But the portion thus supplied was certainly three times as productive as the portion not supplied. My whole net receipts from raspberries amounted to $267. The plants were now well rooted, and were in prime bearing condition. Since this, I have quadrupled my facilities for applying the liquid manure. A large hogshead has been mounted on low wheels, the rims of which are four inches wide, so as to prevent them sinking into the ground, the whole being constructed to weigh as little as possible. The sprinkling apparatus will drench one or two rows at a time, as may be desired. The driver rides on the cart, and by raising or lowering a valve, lets on or shuts off the flow of liquor at his pleasure. Having been used on the raspberries for several years, I can testify to the extraordinary value of this mode of applying manure. It stimulates an astonishing growth of canes, increases the quantity of fruit, while it secures the grand desideratum, a prodigious enlarge-ment in the size of the berries. I find by inquiry among my neigh-bors that none of them get so high prices as myself. Every crop has been growing more profitable than the preceding one; and it may be set down that an acre of raspberries, treated and attended to as they ought to be, will realize a net profit of $200 annually.

The Lawtons were this year to come into stronger bearing. Par-ties in New York and Philadelphia had agreed to take all my crop, and guarantee me twenty-five cents a quart. One speculator came to my house and offered $200 for the crop, before the berries were ripe. I should have accepted the offer, thinking that was money enough to make from one acre, had not my obligation to send the fruit to other parties interfered with a sale. But I made out a trifle better, as the quantity marketed amounted to 896 quarts, which netted me $206.08. In addition to this, the sales of plants amounted to $101. As the market price for plants was falling, I was not anxious to multiply them to the injury of the fruit; hence many suckers were cut down outside of the rows, so as to throw the whole energy of the roots into the berries; and I think the result justified this course. The demand for the fruit was so great, that I could have readily sold four times as much at the same price. As the season for the blackberries closed, all the stray fruit was gathered and converted into an admirable wine. Some seventy bottles were made for home use; and when a year old, I discovered that it was of ready sale at half a dollar per bottle. Since then we have made a barrel of wine annually; and when old enough, all not needed for domestic purposes is sold at $2 per gallon. It is a small item of our general income, but quite sufficient to show that vast profit may be made by any person going largely into the busi-ness of manufacturing blackberry-wine.

We raised nothing of value among the blackberries this year. The growth of new wood had been so luxuriant, that the ground between the rows was too much shaded to permit other plants to mature. In some places, the huge canes, throwing out branches six to seven feet long, had interlocked with each other from row to row, and were cut away, to enable the cultivator and weeder to pass along between them, and thenceforward this acre was given up entirely to the blackberries. As the roots wandered. away for twenty or thirty feet in search of nourishment, they acquired new power to force up stronger and more numerous canes. Many of these came up profusely in a direct line with the original plants. When not standing too close together, they were carefully preserved, when of vigorous growth; but the feeble ones were taken up and sold. Thus, in a few years, a row which had been originally set with plants eight feet apart became a compact hedge, and an acre supporting full six times as many bearing canes as when first planted. Hence the crop of fruit should increase annually. It will continue to do so, if not more than three vigorous canes are allowed to grow in one cluster; if the canes are cut down in July to three or four feet high; if the branches are cut back to a foot in length; if the growth of all suckers between the rows is thoroughly stopped by treating them the same as weeds; if the old-bearing wood is nicely taken out at the close of every season; and, finally, if the plants are bountifully supplied with manure. From long experience with this admirable fruit, I lay it down as a rule that every single condition above stated must be complied with, if the grower expects abundant crops of the very finest fruit. Observe them, and the result is certain; neglect them, and the reward will be inferior fruit, to sell at inferior prices.

To the Lawtons succeeded the peaches, now their first bearing year. We had protected them for three seasons from the fly by keeping the butts well tarred, and they were now about to give some return for this careful but unexpensive oversight. Some few of them produced no fruit whatever, but the majority made a respectable show. I went over the orchard myself, examining each tree with the utmost care, and removed every peach of inferior size, as well as thinning out even good ones which happened to be too much crowded together. Being of the earlier sorts, they came into market in advance of a glut; and though the money-pressure in the cities was now about culminating in the memorable explosion of September, yet there was still money enough left in the pockets of the multitude to pay good prices for peaches. It is with fruit as it is with rum – men are never too poor to buy both. My 804 trees produced me $208 clear of expenses, with a pretty sure prospect of doing much better hereafter. I had learned from experience that a shrewd grower need not be apprehensive of a glut; and that if panics palsied, or a general insolvency desolated the cities, they still contrived to hold as much money as before. Credit might disappear, but the money remained; and the industrious tiller of the soil was sure to get his full share of the general fund which survives even the worst convulsion.

My acre of tomatoes netted me this year $192, my pork $61, my potatoes $40, and the calf $3. Thus, as my grounds became charged with manure, – as I restored to it the waste occasioned by the crops that were removed from it, and even more than that waste, – so my crops increased in value. It was thus demonstrable that manuring would pay. On the clover-field the most signal evidence of this was apparent. After each cutting of clover had been taken to the barn-yard, the liquor-cart distributed over the newly mown sod a copious supply of liquid manure, thus regularly restoring to the earth an equivalent for the crop removed. It was most instructive to see how immediately after each application the well-rooted clover shot up into luxuriant growth. I have thus mowed it three times in a season, and can readily believe that in the moister climate of England and Flanders as many as six crops are annually taken from grass lands thus treated with liquid manure. Indeed, I am inclined to believe that there is no reasonable limit to the yield of an acre of ground which is constantly and heavily manured, and cultivated by one who thoroughly understands his art.

Three years' experience of profit and loss is quite sufficient for the purposes of this volume. It has satisfied me, as it should satisfy others, that Ten Acres are Enough. I give the following recapitulation for convenience of reference:

Expenses for three years.	1855.	1856.	1857.
Manures of various kinds	$268.00	$346.00	$358.06
Wages and labor	102.00	238.00	244.00
Feed for stock..	28.00	79.30	103.00
Stakes and twine for Lawtons		7.00	8.00
Garden and other seeds	8.00	13.00	
Cabbage and tomato plants	30.00		
Lumber, nails, and sundries		14.50	81.00
Loss on cow	7.00		
Cost of pigs	12.00	12.00	12.00
	$455.00	$709.80	$806.06

Receipts for three years.	1855.	1856.	1857.
Strawberries, 6 acres		$857.60	$903.00
Lawton plants sold	$460.00	213.50	101.00
Tomatoes, 1 acre	120.00	190.00	192.00
Garden, including ten peach-trees	80.00	121.00	152.00
Cabbages	82.00	70.20	
Raspberries, 2 acres		38.72	267.00
Lawtons, 1 acre		159.84	206.08

This result may surprise many not conversant with the profits which are constantly being realized from small farms. But rejecting the income from the sale of plants, the pigs, and the calf, as exceptional things, and the profit of the nine acres for the first year will be found to be nothing per acre, for the second year, $83.50, and for the third, $129.10. But there are obvious reasons why this should be so. The ground was crowded to its utmost capacity with those plants only which yielded the very highest rate of profit, and for which there was an unfailing demand. In addition to this, it was cultivated with the most unflagging industry and care. Besides using the contents of more than one barnyard upon it, I literally manured it with brains. My whole mind and energies were devoted to improving and attending to it. No city business was ever more industriously or intelligently supervised than this. But if the reward was ample, it was no greater than others all around me were annually realizing, the only difference being that they cultivated more ground. While they diffused their labor over twenty acres, I concentrated mine on ten. Yet, having only half as much ground to work over, I realized as large a profit as the average of them all. Concentrated labor and manuring thus brought the return which is always realized from them when intelligently combined.

Pork	49.00	58.00	61.00
Potatoes		24.00	40.00
Calf		2.00	3.00
Peaches, 804 trees, first bearing year.			208.00
	$791.00	1,734.86	2,133.08
Expenses as above stated	455.00	709.80	806.06
Annual profit	$336.00	1,025.06	1,327.02

For six years since 1857 I have continued to cultivate this little farm. Sometimes an unpropitious season has cut down my profits to a low figure, but I have never lost money on the year's business. Now and then a crop or two has utterly failed, as some seasons are too dry, and others are too wet. But among the variety cultivated some are sure to succeed. Only once or twice have I failed to invest a few hundred dollars at the year's end. All other business has been studiously avoided. I have spent considerable money in adding to the convenience of my dwelling, and the extent of my outbuildings; among the latter is a little shop furnished with more tools than are generally to be found upon a farm, which save me many dollars in a year, and many errands to the carpenter and wheelwright. The marriage of my daughter Kate called for a genteel outfit, which she received without occasioning me any inconvenience. I buy nothing on credit, and for more than ten years have had no occasion to give a note. If at the year's end we are found to owe any thing at the stores, it is promptly paid. As means increased, my family has lived more expensively, though I think not any more comfortably. I lie down peacefully at night, thinking that I do not deserve more than others, but thankful that God has given me more. I rise in the morning with an appetite for labor as keen as that for breakfast. But others can succeed as well as myself. Capital or no capital, the proper industry and determination will certainly be rewarded by success.

CHAPTER XXI.
A BARNYARD MANUFACTORY –
LAND ENOUGH – FAITH IN MANURE.

As previously stated, there is no successful farming without a liberal expenditure for manure. I had proved that high manuring would pay, and while anxious to increase the quantity, was desirous of reducing the money-cost. I continued every season to scour the neighborhood for leaves, and to gather up every available material for the barnyard. But in addition to all this, in October and November of my fourth year I purchased twenty heifers which would calve in the spring, intending to feed them through the winter, and then sell as soon as they had calved. My idea was, that they could be sold for a profit large enough to cover the cost of keeping them, thus having the manure all clear. I consulted many persons versed in this business, farmers, butchers, and others, before venturing on it, as it was a good deal out of my usual line of operations. I also consulted all my files of agricultural papers, where I found set forth a multitude of experiences on the subject, the most of which led me to conclude that it would be safe to try the experiment. There seemed to be but little danger of loss, even if nothing were made, while it was quite certain a good deal of knowledge would be gained.

I accordingly had a rough shed built, large enough to contain twenty cows, with an entry in front of them and a large feed-room at one end. Then mangers were provided, and a plank gutter laid just back of where the cows would stand, into which all the droppings would fall, and down which the water would run into a wide earthen pipe which emptied into the cistern in the barnyard. Here the cows stood in a row, never being allowed to go out, except an hour or two at noon when the weather was fine. I agreed with Dick to take entire charge of the feeding and watering them, for the consideration of $30 extra. I bought the cornstalks from some twenty acres near me, at $3 an acre, and these were delivered from time to time as they were needed, there not being room on the premises for so large a quantity at once. I had provided a superior cutter, with which Dick cut up the stalks and blades, reducing them to pieces a half-inch long, and he then put them into a hogshead of water, where they remained a day and night to soak. Thence they were transferred to a steaming apparatus, constructed expressly for this purpose, where they were made perfectly soft. Corn meal, bran, and various kinds of ground feed were mixed in and steamed with the cut stalks, a sprinkling of salt being added. A day's feed for the whole twenty was cooked at one operation. This preparation came out soft and palatable, and the cows took to it greedily. The ground feed was varied during the season, and occasionally a few turnips, parsnips, and cabbages were cooked up to increase the variety. I had no hay to give them, and they got none.

But on the other hand, Dick gave them four good strong messes every day, that at night being a very heavy one. He said they throve as well as any cattle he had ever seen. The gutter behind them was cleaned out twice a day and sprinkled with plaster, thus keeping the place always clean and sweet. In fact, I made cleanliness the order of the day throughout the entire barnyard. The manure was thrown directly from the gutter into a wheelbarrow having a thick layer of leaves spread over its bottom, and then emptied in a heap under the manure shed. As the cows were also littered with leaves, these, when too foul for longer use, were taken to the same heap. Others were added, with cornstalks in occasional layers; and as each layer was deposited, the whole heap was saturated with liquor from the cistern. I do not think a better lot of common barnyard manure has ever been manufactured. Dick entered into the spirit of the experiment, and carried it through without once faltering.

As soon as the cows began dropping their calves in the spring, I advertised them, and plenty of purchasers appeared, as a choice out of twenty was of some value. They had cost me $22 each. I had kept them an average of one hundred and forty days for each cow, at a cost of six cents per day for each, or $8.40, making with the first cost $30.40 per cow, or $608 for the whole. To this was added $60 for cornstalks and $40 for Dick, making a grand total of $708. I sold them at an average of $35.50, and thus realized $710, or a cash profit of $2. Instead of paying Dick $30 for his trouble, I told the fellow that as he had performed his duty so satisfactorily, he should have $40. This little voluntary contribution so gratified him, that I feel assured its value has been refunded to me fourfold, by his subsequent attention as a professor of the art and mystery of manufacturing manure.

Thus I made $2 in cash by the operation, besides having a great quantity of cornstalks left over, and a pile of manure certainly as ample as any for which I had paid $250. Moreover, it was on my own premises; it had been most carefully attended to during the whole process of manufacture; I knew what it was composed of, and that the seeds of noxious weeds could not have been added to it. All these facts gave it value over the chance lots which farmers are often compelled to purchase, and from which their fields are many times sowed with thousands of weeds. Here was a clear saving of $250 added to my profits.

The result was so encouraging, that I have continued the practice of thus feeding cattle during the winter from that day to this, increasing the number, however, to twenty-five. I find no difficulty in making sales in the spring. Sometimes I have lost a few dollars on a winter's operations, sometimes made a little profit, and sometimes come out just even. On the run of four years there has been no profit beyond the manure; but that much is all clear. Thus the winter, instead of being a season of suspended profits as formerly, is now one of positive gain. The operation of thus feeding cattle is certainly attended with trouble. But once provided with all the conveniences for carrying it on, it is not only simple and easy, but becomes even interesting. No one who has not tried it in a careful, methodical way, can have any idea of the rapidity with which the manure heaps grow, nor the size to which they ultimately attain. My neighbors having long since ceased to be amused at what they facetiously called the novelty of my operations, did not venture to ridicule even this. On the contrary, they rather approved of it, though not one of them could tell how much it cost to keep a cow per week. But I impute no part of my success to their approval. The practice is intrinsically a good one, and only needs being carried on properly to make it pay.

Let me add, that there is a very cheap and convenient mode of covering manure from the weather, which I have constantly practised, thus avoiding the east of building sheds. I took inch boards which were sixteen feet long, and sawed them in half, making two lengths, each eight feet. The boards were as wide as could be had, say twenty inches. Battens were then nailed across each end and the centre, to prevent warping. Then to each end a board of equal width, and five feet long, was secured by strap hinges. The manure heap was then built up, say five feet high, and eight wide at the top. When thus finished, one of the boards was placed across the top; the ends being hinged, fell down over the sides of the heap, and touched the ground. Beginning at one end of the heap, the hinged boards were laid on until they reached to the other end. Thus the entire heap, except the ends, was complete-ly protected from the weather. The ends were covered with loose boards. Whenever rain was coming on, and it was thought the heap needed water to prevent fire-fanging, this portable shed was lifted off in five minutes. After receiving a good soaking, the shed was in five more minutes replaced on the heap; and when no composting was going on, the boards were simply stowed away in some by-place until again wanted. To those who believe in the value of housing manure, but who cannot afford to erect buildings for the purpose, these porta-ble sheds will be found, for §10, to be as effectual as a building costing §60, while at the same time they do not occupy any useful ground.

I will not say that ten acres in New Jersey can be made to pro-duce more money than ten acres located elsewhere, within reach of the great city markets. Without doubt, the productiveness of either tract will be in exact proportion to the care and skill of cultivation, and the thoroughness of manuring. In either case, it is utter folly for a man to attempt the cultivation of more land than he can manage thoroughly. The chances are then invariably against him. I consider the real office of the ground to be merely that of holding a plant in an erect position, while you feed the roots. But it is nevertheless remarkable that the census tables show that the product of New Jersey per acre, when the whole area of the State is taken into account, is considerably greater than in any of the adjoining States. The product per acre, in some of the fruit-growing counties nearest the two great cities, is even more remarkable. The average cash value of the products of all our market gardens is $20 annually, while that of the gardens in New York and Pennsylvania is only $5 each. Of our orchards it is $25, while in New York it is only $10, and in Pennsylvania only $5. The value of agricultural implements and machinery is relatively far greater than in either of these empire States. Nothing short of a superior productiveness for truck and fruit, in the soil of New Jersey, can account for such results.

A farmer in my neighborhood sold from forty early apple-trees, occupying about one acre of land, 400 baskets of fruit, which yielded, after deducting expenses, and ten per cent. commission for selling, $241.50. I have known pears to be sold at from $3 to $5 per basket, and in smaller quantities at $2 a half-peck; and three cherry-trees, of the early Richmond variety, yielded $30 worth of fruit. Peach- trees, when protected from the worm, will bear luxuriantly for twenty years.

I know a small farmer, with six acres of rhubarb, who has realized $600 dollars annually from it. Another has twenty acres of asparagus, from which he realizes $600 per week during the season for cutting. Besides, it grows an acre of common gooseberries, from which his annual profit is $200. I have known another to sell $500 worth of tomatoes from a single acre, besides having many bushels for the hog-pen. I could name owners of very small tracts who are doing well in the same business. Asparagus, strawberries, raspberries, blackberries, currants, grapes, and gooseberries, grow to perfection, and yield enormous returns when properly attended to, far surpassing any thing ever obtained from the heavier staple crops, such as grain, grass, and stock.

But it is to be noticed that the greatest profit per acre is almost invariably realized by those who have very small farms. The less they have, the more thoroughly is it cultivated; while the few who have sufficient faith in manure, and who thus convert their entire holding into a garden, realize twice or thrice as much per annum as they had paid for the land. I knew a striking illustration of the value of this faith in manure. It is in the person of a Jerseyman who began, twenty-five years ago, upon a single acre of rented land, with a capital of only $50, borrowed from a sister who had saved it from her earnings as a dairymaid. This man regarded the earth as of no practical use except to receive and hold manure; and his idea was, that if he crowded it full enough, every rain would extract from it, and convey directly to the roots of the plants, the liquid nutriment which gives to all vegetation such amazing vigor. Thus, the solids, if in sufficient supply, would be sure to furnish the liquids, on which he knew he could rely. Though full of original and practical ideas, this was his absorbing one ; and he pursued it with an energy of purpose and a liberality of expenditure that surprised the population of an entire township.

In spite of the disadvantages attendant on a neglected education, the force of this man's strong natural sense carried him forward with astonishing rapidity. True to his faith in manure, he bought and manufactured to an extent far exceeding all his neighbors. He soon obtained possession of a small farm, with ample time allowed for payment; for his industry and skill established a character, and character served for capital. In a few years he monopolized the contents of all the pig-pens in the city near which he resided, all that was produced by the slaughter-houses, all the lime from the gas-works, all the spent bark from the tanneries, and every tub of night-soil which came from the water-closets of a large population. He created a demand for manure so general, that the streets were traversed by men and boys with carts and handbarrows, who daily picked up the droppings from the numerous livestock which passed over the roads, and piled them snugly in fence-corners, composting them with leaves and rubbish, knowing that the great manure king would take them all. The quantity thus collected by these industrious scavengers was very large. In addition to all this, he purchased cargoes of marl, charcoal cinders from the pines, guano, and slooploads of manures from the city. The world within his reach seemed unable to supply his vast demand.

His cash outlay for these fertilizers was of course enormous, and has amounted to thousands of dollars per annum. It has been constantly increasing, and grows even as I write. But his faith in manure was accompanied by works. What he thus collected at so great a cost, was applied with singular shrewdness to the production of fruits and vegetables for the great city markets. His fields rewarded him in proportion as he enriched them. His neighbors, who, for miles around, had been astonished and incredulous at his unprecedented outlay for manure, were in turn astonished at the extraordinary quantities of fruit and truck which he dispatched to market. As he went early and largely into the growing of rhubarb, when all others were too timid even to touch it, so for years he was the only man who sent tons of it to market during a long period in which it paid extravagant profits. By skilfully regulating his crops, he secured an uninterrupted succession during the entire season; so that from the earliest to the latest period of the year he was constantly receiving large cash returns. His wagons have sometimes loaded an entire steamboat, sometimes an entire train upon the railroad. By growing asparagus, he has realized great profits. For years he commanded the Baltimore markets with his strawberries, while various other large towns depended on him for their supplies. I have been upon his thirty acres of this fruit during the height of the season, when fifty pickers were at work on ground which wore a tinge of luscious scarlet under the astonishing profusion of the crop; while thousands of quarts, under adjacent sheds, were in process of being boxed for market. Of this fruit he has sent ninety bushels to market in a single day, distributed $300 in a week among his pickers, while in the boxes to contain them his investment is $1,500. On strawberries alone this man could have grown rich.

But they are scarcely a tithe of what he has produced. Raspberries, blackberries, and all the smaller fruits have been cultivated quite as extensively. The same courageous intelligence which led him to outstrip all competitors in the application of manure, kept him awake to every improvement in fruit or vegetable as it came before the public. He not only procured the best of every kind, but bought them early, no matter how extravagant the price. Thus keeping in advance of all others, so his profits have exceeded theirs. More than once he has been cheated by the purchase of novelties of this kind, besides losing time and money in cultivating them long enough to prove the cheat; but these losses have been but as dust in the great balance of his profit.

As may be supposed, such a man could not fail to become rich. From his humble beginning of a single acre he has gone on adding farm to farm, house to house, and lot to lot, and is ever on hand to purchase more. His passion is to own land. But even so thorough a farmer as he may in the end acquire too much to be profitable.

The example thus set has had a marked influence on the population of entire townships. Men who at first, and who even for years, were incredulous of the propriety of using such vast quantities of manure, at length became converts to the example. High farming thus came extensively into vogue. Meantime the facilities for getting to market Avere being constantly multiplied. New fertilizers were introduced and kept for sale in all the country towns, the facility for obtaining them thus inducing a general consumption. As crops increased, so the great cities grew in size, the number of mouths to be fed enlarging with the supply of food. Under the pressure of all these several inducements, fruit and truck have been produced in quantities that cannot be estimated.

The first great impulse to its enlarged production in the neighborhood where the enterprising consumer of manure resided, to whom reference has been made, was the result of his example. His great success removed all doubt and disarmed all opposition. But even his was not achieved without unremitting industry and intelligent application of the mind. Neither his hands nor the manure did every thing. But manure lay at the foundation of the edifice: without it he would have toiled in vain to build up an ample fortune from the humblest of beginnings. As he succeeded, so let others take counsel, and have faith in manure.

CHAPTER XXII.
PROFITS OF FRUIT-GROWING – THE TRADE IN BERRIES.

IT cannot be supposed that agriculture is always a successful pursuit. On the contrary, we know it many times to be the reverse. But when one looks carefully into that branch of it which embraces fruits, especially the smaller kinds, the evidences in its favor as a money-making business multiply as we proceed. The reader must have some knowledge of the prodigious profits realized a few years since by the peachgrowers in Delaware, where 800 acres were cultivated in that fruit by a single individual. At one time he was compelled to charter several steamboats during the entire season, to convey his thousands of baskets to market. From only 70 acres the owner has realized a net profit of $12,000, in one season. The instance of my relative in Ohio, mentioned in an earlier chapter, affords another illustration of what a very small orchard can be made to yield. I have known single peach-trees in gardens, in seasons when the general crop was short, producing as much as $20 each. Those who buy single peaches at the street corners in our cities, one or two for a dime, can readily understand these figures, I could point out a garden belonging to a widow, containing twelve plum-trees, from which she regularly receives $60 every year, and sometimes even more. Grapes are never so abundant in market as to reduce the price below the point of profit.

The prices paid for pears are such as to seem absurdly high. But even when rebellion had most depressed the market, I knew a single tree to net $23 to the owner. Another grower, from three trees, annually receives $60. A citizen of New York is the owner of three pear-trees which have yielded eleven barrels, and produced $137. There is another tree in that State, seventy years old, from which, in that period, $3,750 worth of pears have been sold – enough to pay for a farm. A young orchard of four hundred trees, some eight years after planting, at two years' crops yielded the owner $1,450. An acre of the best pear-trees, well managed, will produce more profit than a five-hundred-acre farm, without a twentieth of the care or capital.

But examples almost without number may be given, where apple-trees also have yielded from five to ten dollars a year in fruit, and many instances in which twenty or thirty dollars have been obtained. If one tree of the Rhode Island Greening will afford forty bushels of fruit, at a quarter of a dollar per bushel, which has often occurred, forty such trees on an acre would yield a crop worth four hundred dollars. But taking one quarter of this amount as a low average for all seasons, and with imperfect cultivation, one hundred dollars will still be equal to the interest on fifteen hundred per acre. Now, this estimate is based upon the price of good winter apples for the past thirty years, in one of our most productive districts; let a similar estimate be made with fruits rarer and of a more delicate character. Apricots and the finer varieties of the plum are often sold for three to six dollars per bushel, and the best early peaches from one to three dollars. An acquaintance received eight dollars for a crop grown on two fine young cherry-trees, and twenty-four dollars from four young peach-trees of only four years' growth from the bud. In Western New York, single trees of the Doyenne or Virgalieu pear have often afforded a return of twenty dollars or more, after being sent hundreds of miles to market.

These standard fruits, requiring several years to come into bearing, are too slow for the majority of cultivators, who, like myself, need something which will pay in a year or two. The whole berry family is pre-eminently adapted to meet this demand for immediate profit. Happily for the multitude engaged in its propagation, the business cannot be overdone. Could an exact calculation be made of the money expended in the city of New York merely for the small fruits, the amount would be so enormous as to be scarcely credible, and would go far to prove the immense wealth which actually exists, in spite of the fact that thousands are suffering all the stings of poverty. Take the strawberry as a faint index of the large sums of money that are annually laid out in the different varieties of fruit.

One of the most ephemeral of all fruits, only lasting its brief month, the strawberry nevertheless plays no insignificant part in the *role* of our early summer business. In fact, this little berry may be said to be the prime favorite of the season. Of a delicious flavor, with just sufficient of tartness to render it agreeable, it commends itself to the faste of young and old; while its cooling properties render it highly beneficial, in a hygienic point of view, during the early heats of the dog-days. Then its cheapness places it within the reach of the poorest. It is alike welcome to the schoolboy who has a few cents of pocket money to invest in such delicacies as schoolboys are wont to indulge in; to the laboring man, after the burden and heat of the day are over; and to the wealthy, who has at his command the means of enjoyment of the most expensive kind.

The first strawberries during the season generally appear at the Broadway saloons about the middle of May, and are sold at the very modest price of fifty cents per pint basket. A placard in the window announces that a plateful, with cream, may be had for a similar small consideration. These early strawberries are from Virginia; but as they are small, with immaturity stamped upon them, it is to be presumed that there is not a very great rush for fifty cents' worths, even by such as feel like boasting that they had eaten strawberries and cream ere the frosts of winter had well disappeared. Soon, however, New Jersey begins to give up her stores of the delicious fruit, and prices fall from fifty to fifteen, from fifteen to six, from six to five, and finally from five to three cents per pint.

Almost the entire early crop of the New York market is grown in New Jersey, and by far the largest quantity brought into the city by any one route reaches New York by the New York and Erie Railroad Company. The berries are conveyed in carts and wagons from the gardens where they are grown to the several railroad stations, whence they find their way to the respective ferries. Great quantities, however, are conveyed in wagons direct to the ferries. Hence it is next to impossible to obtain exact information of the actual quantities brought into the city, and consumed by the inhabitants. All that can be done is to convey an approximate idea of the immense extent of the trade, leaving the reader to imagine what must be the actual quantity, since that of which authentic information can be obtained is so enormous.

The berries are largely shipped from Burlington, Monmouth, and Middlesex counties in New Jersey. Large quantities are also grown in Bergen. The Bergen County Journal says, that from data furnished, it considers 10,000,000 baskets a low estimate of the quantity sent to market in one season from that county alone. This evidently is a mistake, for, after a very close inquiry into the matter, it does not appear that any thing like that quantity has reached New York from all places where the berry is grown. Even supposing that other markets besides that of New York are indicated, the quantity named seems too large for credibility, as having been grown in a single county, however favorable the soil may be to the production of the fruit, and notwithstanding the utmost indefatigability of the growers; and the more so when the Journal adds, " that thousands, perhaps millions of baskets, have rotted on the vines."

The opening of the Northern Railroad of New Jersey to Pier-mont, is another circumstance which has given an impetus to the trade. The opening took place just at the commencement of the season of 1859, – not early enough for the growers to make their ar-rangements for a very large crop, but just in time to enable them to take full advantage of the means of transit over the line, of the then ripening crop. Accordingly, as far as can be ascertained, 400,000 baskets were brought over the new road. This looks well for a com-mencement, and holds out a good promise of an enormous trade in future seasons. The section of country through which the line runs, quietly undulating, is well watered, and admirably adapted to the growth of the strawberry; and as the settlements arc within easy dis-tances of the stations, the fruit can be sent into market fresh picked and sound, retaining its full, rich flavor.

The cultivation of the strawberry is very little attended to on Long Island. On inquiry at the railroad station there, it was found that so small is the quantity brought over by it, that it was not deemed worth while to charge freight for the few parcels carried by travellers. The quantity may be safely set down at 25,000 baskets. No business is done in this fruit over the Hudson or the Harlem and New Haven Railroads.

Besides the railroads, the steamboats bring to market large quantities of the fruit. It is impossible to obtain correct statistical information of the trade from this source. The quantity brought from Keyport, N. J., alone, by two vessels, has been estimated at 1,750,000 baskets.

The following is an epitome of the business done, as far as can be ascertained: –

	Baskets.
Over the New York and Erie Railroad	3,253,407
" " Railroad of Northern New Jersey	400,000
" " Long Island Railroad	25,000
" " Camden and Amboy Railroad	1,100,000
From Keyport, in vessels	1,750,000
" " Hoboken and other places, in wagons	500,000
	7,028,407

Say seven millions of baskets, in round numbers. Of the three and a quarter millions brought over the New York and Erie Railroad, somewhat more than one-half are from Ramsey's and Allendale station, and the remainder from the stations on the Union Railroad and the Piermont branch. Of those brought by the Camden and Amboy Railroad, the great bulk is from Burlington county.

It is difficult to form a correct estimate of the average price at which strawberries sell; but by carefully collating the statements of the principal wholesale dealers, and taking the mean of the several prices, throughout the season, $3 per hundred baskets, by wholesale, seems to be pretty near the mark. From the wholesale dealers the article sometimes changes hands twice, before reaching the consumer, who, taking the average, may be said to have paid 31/2 cents per basket, or $3.50 per hundred. Consequently, it will be seen that the retailer makes but a small profit, especially in eases where the strawberries reach him through the hands of the middle-man, who of course manages to make his share of gain in the transfer. The wholesale dealers generally sell on commission, accounting to the growers for their sales, and reserving ten per cent. for their trouble. The largest quantity sold by any one dealer is about 300,000 baskets. The freight charge over the railroads is 121/2 cents per hundred baskets.

The following figures will show what a conspicuous part this apparently insignificant berry plays in our social economics:

700,000 baskets, at $3.75 per hundred................. $210,000
Profit to the retailers, at 75 cents per hundred........ 45,000
Commission to wholesale dealers, at 10 per cent...21,000
Freight, at 121/2 cents per hundred, all round............8,750

This is only as far as can be ascertained, but there is reason to believe that thousands of baskets of strawberries find their way into the New York markets, of which no account can be obtained, thus tending to swell the enormous expenditure on this almost the smallest of summer fruits.

It is equally difficult to ascertain the quantity of this fruit which pours into Philadelphia also, during the season, but it is probably two-thirds as great as that which goes to New York. There are numerous growers near the former city, who dispatch to it from twenty to sixty bushels each, daily.

An experienced writer on this subject estimates the consumption of strawberries in the four great cities as follows –

New York...54,000 bushels.
Philadelphia...14,000 "
Boston..11,000 "
Cincinnati..14,000 "

This estimate of the consumption of Philadelphia is a very erroneous one, as the consumption must fully equal that of New York. In 1860, no less than 173,500 quarts of strawberries passed through the gate of only one of the numerous gravel turnpikes in New Jersey, on their way to Philadelphia. This is equal to 5,4421/2 bushels, more than one-third of the quantity estimated as above.

He says that 8,000,000 baskets (five to the quart) have been received in New York in a season. He adds, that the crop around these four cities does not exceed 25 to 50 bushels per acre, although instances are reported where 100, and even 130 to 140 bushels have been produced on an acre, or in that proportion. The returns, therefore, vary from $100 to $800 per acre, and the prices range from $1.50 down to 121/2 cents a quart. The former price is readily obtained in Washington at the opening of the season.

He thence argues that in order to supply New York and vicinity with strawberries, about 1,500 acres, of the choicest land is required, and 500 for the other cities named. This he alleges to be at least four times as much land as is either appropriate or necessary for the object, if the nature and cultivation of the strawberry were only as well understood as the raising of corn. He contends that a crop of thirty bushels of strawberries to the acre, is only about proportionate to a corn crop of ten bushels on the same ground. He says that a strawberry plantation is seldom seen without having, after the first year, many more plants upon the ground than can obtain air or light sufficient to fruit well. The consequence is, that all our city markets are mainly supplied with inferior fruit, simply because some of the commonest kinds continue to produce a little stunted, sour fruit, even under the worst treatment. Superior, well-grown fruit will easily produce twice and four times as much to the acre, and will command prices from two to four times larger in the city markets: making the avails and the difference from the same land to be 25 bushels at 121/2 cents a quart, or at least 125 bushels at 25 cents a quart, or $1,000 or $100 an acre. He lays it down that an acre ought to be made to yield 125 bushels, and that no grower should be satisfied with less.

That this yield and these profits can be realized, there are. numerous evidences. Small plots of ground, thoroughly cultivated, have yielded even a double ratio. One grower in Connecticut realized $215 from strawberries raised on twenty-five rods of ground, or at the rate of $1,300 per acre. A citizen of Maine has raised them, on a small lot, at the rate of 300 bushels an acre. Another in New Jersey cleared $1,100 from three acres, and one of the agricultural societies in that State awarded the strawberry premium to a gentleman whose ground produced them at the rate of $1,222 an acre, clear profit. I have seen a crop ripening on three acres for which the owner was offered $800 as it stood, the buyer to pick and take it away at his own expense. The offer was declined, and the owner realized $1,300 clear. Mr. Fuller, of Brooklyn, has grown at the rate of 600 bushels per acre, on a small plot of the Bartlett; and by the same mode of treatment, 400 of the Triomphe de Gand.

All these returns are unquestionably the effects of high culture. Those who fail to practise it, also fail to realize such returns. The slovenly cultivator complains that his strawberries run out. But this is because he permits the weeds and grass to run in and occupy the ground. The plant has no inherent tendency to degenerate. For the last few years, immense demand has existed for Wilson's Albany Seedling. Those at all conversant with the subject, know that plenty of room is requisite to get the greatest quantity of runners from a given number of plants – the sale being perfectly sure, all dealers give this room; the consequence is, while the plants are worth say $10 per 1000, all are fine large plants, and give a fair crop, even the first year after planting. Such plants tell their own story, and the demand continues. In a short time, prices come down; and the supply increasing beyond demand, the dealer no longer thinks it worth while to give this room expressly for the growth of plants: the beds take care of themselves, hence bear but little, and the plants furnished are always weak and spindling. These require the second year to fruit; perhaps, in the interim, new kinds are pressed into notice, and from the old beds it becomes more and more difficult to obtain strong plants, until the cry is raised that the once celebrated strawberry has run its race. Now, the question is, whether the same kinds under the same circumstances, that is, strong runners from strong old plants, in good soil and plenty of room, will not continue to be productive.

As this is not designed to be a treatise on the art of raising straw-berries, so I shall not enlarge upon the subject. Every grower seems to have a method of his own, which he prefers over all others. There are works upon the subject, containing numerous facts with which every careful beginner should make himself familiar. But even in these are to be discovered the most extraordinary collisions of opin-ion, – one, for instance, Recommending generous manuring, an-other insisting that poor ground only should be used, while a third declares that frequent stirring of the soil will of itself insure abun-dant crops. Amid all these antagonisms one great fact stands promi-nently forth, that the strawberry plant will continue to live and pro-duce fruit under every possible variety of treatment; while another is equally conspicuous, that the better the treatment the better the return. It would be presumptuous in a novice like me to undertake to reconcile these unaccountable discrepancies of the great strawberry doctors of the country. But I have learned enough to be satisfied that *soil* has much to do in the successful cultivation of this fruit. A variety which flourishes in one soil will be almost barren in another. Hence, in the . hands of one grower it proves a great prize, but in those of an-other it is comparatively worthless. Without doubt it is to this cause that much of the diversity of opinion as to certain varieties, as well as to the mode of culture, is to be attributed.

Neither will I undertake to decide what sorts, among the cloud of new aspirants for public patronage which are annually coming into notice, are to be adopted as the best. One is in danger of being con-fused by going largely into the cultivation of a multitude of varieties. Having secured a supply of a few which he has proved to be conge-nial with his particular soil, he should adhere to them. Small trials of the new varieties may be safely made, but wholesale substitutions are many times disastrous undertakings. Having found out such as suit my soil, I am content to keep them. The Albany seedling grows upon it with unsurpassed luxuriance, and I shall probably never abandon it. Meantime I have tried the Bartlett, and found it a rampant and hardy grower, bearing the most abundant crops of luscious fruit. So I find

McAvoy's Superior to be a beautiful berry, and a vigorous runner. In my soil the Triomphe de Gand does not realize the extravagant promise of fruitfulness which heralded its introduction to public notice. My neighbors also complain of it in the same way. But for my own family consumption, I prefer it to any strawberry I have ever eaten. The flavor is rich and luscious beyond description, while the crisp seeds crackle between your impatient grinders with reverberations loud enough to penetrate the utmost depths of a hungry stomach. So long as my vines continue to produce only one- fourth as much as others, I shall continue to grow this unsurpassable variety. It sends off runners in amazing abundance. When grown in stools, with the runners clipped off weekly, it bears profusely of enormous fruit; and this method, I am inclined to believe, is the true corrective of all unfriendly elements in the soil. In addition to these, I have, in common with " all the world and the rest of mankind," the Tribune strawberries, now growing finely in pots, and carefully housed for crop next summer. Having seen them in fruit, and having also entire confidence that the association by whom they are distributed would no more spread abroad a worthless article than they would circulate a vicious sheet, so I regard the propagation of these three plants as the beginning of a new era in the history of strawberry culture.

I have very little doubt that there are specific manures for the strawberry, and one of them will probably be found in Baugh's Rawbone Superphosphate of Lime. This article is manufactured in Philadelphia, and is made of raw, unburnt bones, which in their raw state contain one-third of animal matter, and combines ammonia and phosphoric acid in the proper proportions for stimulating and nourishing vegetable growth. I have used it as freely as I could afford to, on turnips, celery, and strawberries. On the two former its effect was very decidedly favorable. My celery uniformly exceeds that of my neighbors, both in size, crispness, and flavor, and consequently commands a higher price. But its effect on strawberries has been perfectly marvellous. On some of them the superphosphate was scattered on both sides of the row, whence, by repeated hoeing and raking, with the aid of sundry rains, its finer particles found their way to the roots. The result has been a robust growth of the plants, such as cannot be seen on any other part of my ground. They hold up their heads, their leaves and fruit-stalks some inches higher than any others, while their whole appearance indicates that they have been fed with a more congenial fertilizer than usual. Many of them have put forth double crowns, showing that they are prepared to furnish twice the ordinary quantity of fruit. So impressed am I with the superior value of this fertilizer, that I have, this autumn of 1863, manured as many rows as I could, and shall hereafter substitute it wholly for all barnyard manure. It is applied with the utmost facility, it contains the seeds of no pestiferous weeds, and its virtues are so highly concentrated that a small amount manures a large surface. It is quite possible that it may not do so well on some soils as others, but no farmer can be sure of this until he has made the trial. Hence, as that can be made with a single bag, the sooner it is undertaken the better it will be for those to whose soil it may be found congenial.

Thousands of dollars' worth of the common wild blackberry are annually taken to the cities and sold. For these berries the price has, within a few years, actually risen one-half. The traffic in them on some railroads is immense, especially on those leading into Philadelphia from Delaware. Millions of quarts are annually sold in New York and Cincinnati. A single township in New Jersey sells to the amount of §2,000 and one county in Indiana to that of §10,000. The huckleberry trade of New Jersey is also very large. A single buyer in Monmouth county purchases sixty bushels daily during the picking season. All these wild berries are gathered by women and children who, without these crops, would find no other employment. But they grow in every wood and swamp, in every neglected headland, while upon the old fields they enter into full possession. As they cost nothing but the labor of gathering them, so they are the bountiful means of drawing thousands of dollars into the pockets of the industrious poor. The cranberry swamps of New Jersey are as celebrated for the abundance of their products as their owners have been for permitting them to become the prey of all who choose to strip them of their fruit.

Thus the demand for even the wild berries continues to enlarge. Hence there must be sure sale for those of a superior quality. In fact, the cultivation of fruit is yet in a state of infancy; it is just beginning to assume the character its merits deserve. Probably more trees have been raised, more orchards planted, within the past ten or twelve years than in all previous time. Within a few years past it has received an unusual degree of attention. Plantations of all sorts, orchards, gardens, and nurseries, have increased in number and extent to a degree quite unprecedented; not in one section or locality, but from the extreme north to the southern limits of the fruit-growing region. Horticultural societies have been organized in all parts; while exhibitions, and National, State, and local Conventions of fruitgrowers have been held to discuss the merits of fruits, and other kindred topics, until it has become the desire of almost every man, whether he live in town or country, to enjoy fine fruits, to provide them for his family, and, if possible, to cultivate the trees in his own garden with his own hands.

There are now single nurseries in this country where a million fruit-trees are advertised for sale. If every hundred-acre farm were to receive fifty trees, all the nurseries would be swept bare in a single year. The States east of, and contiguous to, the Mississippi river, would require ten thousand acres of land for three hundred years, to plant ten acres of fruit-trees on every hundred-acre farm in this portion of the Union: and this estimate is based on the supposition that all the trees planted do well, and flourish. If only a fifth of them perish, then two thousand years would be required, at the present rate of supply, to furnish the above-named quantity of orchard for every farm. Some nurseries already cover 300 to 500 acres, but even these go but a short way in supplying the immense demand for fruit-trees. How absurd, then, in the face of such an array of facts as this, the idea that our markets are to be surfeited with fruit ! Thousands of acres of peach-trees, bending under their heavy crops, are still needed for the consumption of but one city; and broad fifty-acre fields reddened with enormous products, may yet send with profit hundreds of bushels of strawberries daily into the other. If, instead of keeping three days, sorts were now added that would keep three months, many times the amount would be needed. But the market would not be confined to large cities. Railroads and steamboats would open new channels of distribution throughout the country for increased supplies. Nor would the business stop here. Large portions of the Eastern Continent would gladly become purchasers as soon as sufficient quantities should create facilities for a reasonable supply. Our best apples are eagerly bought in London and Liverpool, where $9 per barrel is not an unusual price for the best Newtown pippins. And, by being packed in ice, pears gathered early in autumn have been safely sent to Jamaica, and strawberries to Barbadoes. The Baldwin apple has been furnished in good condition in the East Indies two months after it is entirely gone in Boston. The world has never yet been surfeited with fruit.

CHAPTER XXIII.
GENTLEMAN-FARMING –
ESTABLISHING A HOME.

I AM sure I ought not to be considered as belonging to the class of gentlemen farmers. They go into the country because they are rich – I went because I was poor. Yet they have done good service to the public in various ways. They have imported, naturalized, and propagated valuable vegetables and fruits. They have patronized costly labor-saving farm machines and agricultural implements, they have made expensive agricultural experiments, in the benefits of which all cultivators have participated. Especially has this been so in relation to fertilizers, foreign and domestic, natural and artificial. They have improved the breed of domestic cattle, and imported the best blood from abroad, including all the fine-woolled sheep. They have shown us how large crops can be grown, and have otherwise and in various ways radiated good influences around them, and contributed science, dignity, and encouragement to the farmer's vocation.

No one can justly deny the value of their services. Yet it is not by merely cultivating new trees and plants, and exhibiting large vegetables, gigantic apples or pears, corpulative pumpkins, and enormous general crops, that agriculture is to be substantially improved and made profitable to the farmer who depends upon it for a living. Something more than prodigious crops or beautiful fruit is necessary for him. He wants to know the cost of them – to see the balance-sheet in which, while credit is given for the sales of all these fine products, deductions are made for the expenditures rendered necessary to secure them. A tree may produce splendid fruit, but the pears may be few, the apples may be very perishable, and the choice peaches and other trees may bear only every other year, or only once in three or four perhaps, and then die before another crop. The accounts must therefore comprehend several years before the real profits of farming can be truly ascertained.

Herein it is that gentleman-farming is most commonly in fault. The pecuniary results are never either accurately known and stated, or are neglected, because of little consequence to the proprietor. When they happen to be ascertained and divulged, they are often discovered to be far from remunerative. This disregard of cost has brought this genteel agriculture, as it may be called, into disrepute. Common people turn away from it, as inapplicable to the condition of their purses. They think they cannot afford it; and doubtless they are really unable to indulge in this species of agricultural luxury.

It may thus be assumed that this kind of agriculture, so far from being serviceable to many working farmers, is really injurious to them. They confound this uncalculating, heedless practice with book-farming. They believe the conduct of their wealthy neighbors, who follow farming as an amusement, merely relying on their city business for their incomes, to be regulated by the instructions in the agricultural publications of the day. But I fear that this description of literature does not occupy them much. Moreover, it is wisely cautious in its recommendations, as those must be who have witnessed the futility of so many speculations and experiments. High farming is not bottomed on book-learning, if it fails to make suitable deductions for the cost of every operation. In truth, gentleman-farming is too rarely founded on any thing but a full purse, and an ambition to outshine all rivals at a county fair, without much regard to expense. As far as this is true, such agriculture is neither beneficial in a pecuniary view, either to themselves or to the working farmer. The latter finds little in such cultivation that he can copy, because the essential element of expense is left out of the computation. But book-farming ought not to fall under censure because genteel farming happens not to be lucrative.

For the man who can afford to buy almost every thing he needs, and sell very little that he raises, farming is undoubtedly a delightful amusement. For the man who can afford to sell almost every thing he raises, and whose wants are moderate as mine, farming is a lucrative employment. To the oft paraded statistics of premium reports I cannot answer with a sneer. The question is simply this – whether farming, upon the whole, is a profession warranting a certain degree of scientific culture, and giving room for its display – whether it is worthy to enlist the energies and ambition of a young man who has a good life to live, and a career to make? This question may be answered by looking almost anywhere around us. No doubt a farmer should have some practical familiarity with those facts, whether of science or experiment, which have a bearing on his trade. It would be well for him to understand chemistry in its application to farming, yet he should also assiduously gather up those unexplained facts for which even chemistry cannot account.

It would be well for him to know why the johns- wort, the wild carrot, and the Canada thistle thrive so heroically in spite of bad treatment, where are their weak points, where the heel of these Greeks, what degree of heat in the compost pile will destroy the germinating power of seeds, and whether the law of one seed is the law of another seed. He should be a man of business and of some means, for he has his system to decide upon, his labor to engage and direct, his stock and implements to buy, and then his crops to sell, his bills to pay, and his books to balance. Superphosphates certified to by one set of gentlemen-farmers, and the most brilliant eulogies on American farmers, delivered by another set, will not help him much at these things. Money may: indeed every farmer ought to have a little of this commodity to start him fairly.

In almost all locations there are difficulties to encounter. One of these is that of securing efficient laborers. American laborers of the right sort are rarely to be found. American blood is fast, and fast blood is impatient with a hoe among carrots. It is well enough that blood is so fast, and hopes so tall. These tell grandly in certain directions, but they are not available for working over a heap of compost. Farm labor, to be effective, must have the personal oversight of the master. There is breadth and significance in the old saying of Palladius, "If you would push a crop through, look after it yourself."

Another difficulty is the lack of desirable market facilities. The middle-man stands between the producer and the consumer, and monopolizes much of the profit. In this respect farmers might help each other by judicious combination, but they lack coherency as a class. They have too little *esprit du corps*. There is too much of isolation, and isolation will inevitably prey upon the farmer's purse. Then Young America has a growing aversion to manual labor. He is a gentleman; and shall a gentleman take off his coat? He is vain of his culture, and is mortified to find that ordinary sagacity and a rude energy surpass him in success. He learns with pain that knowledge is not confined to books, and that the shrewdness which can mould raw laborers into effective helps, tells more upon the year's profits than the theories of Liebig, or the experiments of Lawes.

But the difficulties thus referred to are many of them gradually disappearing. The labor question, especially, has been wonderfully simplified by the introduction of new and effective implements, which enable the farmer to reduce the number of his hands. But since they do exist, – and I think my representations, though they may seem to show the shady side of the business, will be sustained by the testimony of practical men, – it is best to meet the whole truth in this matter, whatever ugly faces it may wear. No man conquers a difficulty until he sees it plainly. Oaks are fine things, and rivers are fine things; and so are sunsets, and morning-glories, and new-mown hay, and fresh curds, and milch cows. But, after all, a farm, and farming, do not absorb all the romance of life, or all its stateliest heroics. There is width, and beauty, and independence indeed; but there is also sweat, and anxiety about the weather, the crops, and the markets, with horny hands, and sometimes a good deal of hay-dust in the hair. But if a man, as has been said, is thoroughly in earnest; if he have the sagacity to see all over his farm, to systematize his labor, to carry out his plans punctually and thoroughly; if he is not above economics, nor heedless of the teachings of science, nor unobservant of progress otherwise, nor neglectful of the multitude of agricultural lights which shine everywhere around him; let him work, and he will have his reward. But work as he may, it will be impossible to toil harder than thousands in the cities; who, with all their toil of head and hands, end life as poverty-stricken as when they began.

Somehow it happens, that almost every man who has been city-bred feels at times a strong desire to settle down among the trees and green fields, from a vague and undefined belief that the country is the scene where human life attains its highest development. He cherishes a hope, though perhaps a faint one, that he may yet possess a country home, where he may tranquilly pass his latter years, far away from city tumults and trials. This hope is founded on the instinctive desire there is in human nature to possess some portion of the earth's surface. I know that one looks with indescribable interest at an acre of ground which is his own. I am sure that there is something remarkable about my trees. I have a sense of property in every sunset over my own hills, and there is perpetual pleasure in the sight of the glowing landscape at my own door. I have found Ten Acres Enough; and I know well what pleasures, interests, and compensations are to be found in the little affairs of that limited tract. The windows of the snug library, into which I retire in winter, look out across the garden on the blank gable of my barn. When I came here, it was rough and unsightly. But now that homely gable is a blank no longer. Every inch is clustered over with climbing roses, honeysuckles, and variegated ivy, in whose tangled mass of vine and foliage the song-birds build in summer, while to the same annual granary the snowbirds come in flocks to gather seeds in winter. Though I could not aspire to being a gentleman-farmer, seeing that I came to make my fortune, not to spend one, yet I have sought to make farming a sort of social science, in which not only the head and hands could be employed, but the sympathies of the heart enlarged and elevated. In short, to establish a home for the family.

I desire no association with the man or boy who would wantonly kill the birds that sing so cheerfully around our dwellings and our farms: he is fitted for treason and murder. Who among us does not, with the freshness of early morning, call up the memory of the garden of his infancy and childhood; the robin's nest in the old cherry-tree, and the nest of young chirping birds in the currant-bush; the flowers planted by his mother, and nurtured by his sisters? In all our wanderings, the memory of childhood's birds and flowers is associated with that of mother, sister, and our early home. As you would have *your* children intelligent, virtuous, and happy, and their memory, in after-life, of early home a pleasant or repulsive one, so make your farms and your children's home as your business of life, then adorn that business throughout. If you would inspire your own children and your neighbors with the nobleness of your business, then draw about you such an array of beauty as no one but the cultivator of the soil can collect. Let every foot of your farm show the touch of refinement. While you are arranging your fields for convenient and successful cropping, let it be done with order and neatness. While building the fence, let it be beautiful as well as substantial. While arranging your vegetable-gardens and orchards, do not overlook geometrical regularity. Do not, on any account, omit the planting of flowers and the various kinds of fruit-trees.

CHAPTER XXIV.
UNSUCCESSFUL MEN –
REBELLION NOT RUINOUS
TO NORTHERN AGRICULTURE.

LOOKING back upon the incidents of my city life, I confess that increasing years bring with them an increasing respect for those who do not succeed in life, as these words are commonly used. Heaven has been said to be a place for those who have not succeeded upon earth; and it is surely true that celestial graces do not best thrive and bloom in the hot blaze of worldly prosperity. Ill success sometimes arises from superabundance of qualities in themselves good, from a conscience too sensitive, a taste too fastidious, a self-forgetfulness too romantic, a modesty too retiring. I will not go so far as to say, with a living poet, that

"The world knows nothing of its greatest men;"

but there are forms of greatness, or at least excellence, that die and make no sign; there are martyrs who miss the palm, but not the stake; there are heroes without the laurel, and conquerors without the triumph.

It cannot be denied that there is a class of men who never succeed in business. With a fair amount of earnest industry, they are still unable to get on. Bad luck seems to be their fate, and they are perpetually railing at fortune. In this they are not without sympathy. There are hundreds of simple, good- hearted people, who regard them as ill-starred mortals, against whom an inscrutable destiny had set itself, and who are always ready to pity their mischances and help them in their last extremity. But is not that a very foolish philosophy which refers the misfortunes or the prosperity of individuals to preternatural causes, or even natural causes entirely foreign to the persons? Some people, it is true, owe a great deal to accident. Much of their success is due to circumstances not of their own making. So it is with others who suffer disappointment or disaster. But in those cases in which failure or success is certainly dependent on no extraneous agencies, but on one's own means and energies, I am confident that no little of the complaint of our hard lot is misdirected, and that the charity which helps us out of our successive difficulties is misplaced. In plain words, our failures in this or that thing are often attributable to the fact that we engage in enterprises beyond our power. The world is filled with examples of this truth. We see hundreds of men in all professions and callings who never achieve even a decent living. The bar of every city is crowded with them. They swell the ranks of our physicians and theologians, and swarm in the walks of science and literature; in short, they run against and elbow you everywhere. They are the unfortunate people who have mistaken their mission. They are always attempting tasks which they have not the first qualification to perform. Their ambition is forever outrunning their capabilities. They fancy that to call themselves lawyers, doctors, divines, or the like, is to be what they are styled. Their signs are stuck thickly on doors and shutters all over the city, but they are without honor or employment. Of course they never prosper. They have no fitness for their vocation, no practical skill, no natural talent, and hence they fail.

But both they and society are losers by this. There is so much real ability for something useful that is thus sunk and wasted. The community is encumbered with a host of very incompetent barristers, preachers, physicians, writers, merchants, and so forth, and is deprived of as many good mechanics, and farmers, and laborers. What a pity it is that men will not be content to choose their pursuits according to their abilities. To encourage them to persist in any business for which they are not suited, and in which they never can obtain fortune or credit, is really unkind. It would be much less cruel to let them early feel the inconveniences of following a calling for which they are unfit, and go into one for which nature may have given them the requisite aptitude and powers.

But, in the ordering of a good Providence, failure in one pursuit does not imply failure in the next. I know and have proved this. The motto should be to keep moving, to try it again. Try it a hundred times, if you do not earlier succeed, and all the while be studying to see if you have not failed through some negligence and oversight of your own. Do not throw down your oars and drift stern foremost, because the tide happens to be against you. The tide does not always run the same way. Never anchor because the wind does not happen to be fair. Beat to windward, and gain all you can until it changes. If you get to the bottom of the wheel, hold on – never think of letting go. Let it move which way it will, you are sure to go up.

If in debt, do not let time wear off the edge of the obligation. Economize, work harder, spend less, and hurry out. If misfortune should overtake you, do not sit down and mope, and let her walk over you. Put on more steam, drive ahead, and get out of the way. If you meet obstacles in your path, climb over, dig under, or go around them – never turn back. If the day be stormy, you cannot mend matters by whining and complaining. Be good natured, take it easy, for assuredly the sun will shine to-morrow.

If you lose money on a promising speculation, never think of collecting a coroner's inquest about your dead body. Do not put on long a face because money is not so plentiful as usual – it will not add a single dollar to the circulating medium. Preserve your good-humor, for there is more health in a single hearty laugh than in a dozen glasses of rum. Be happy, and impart happiness to others. Look aloft, and trust in God. Be prudent as you please, but do not bleach out your hair, and pucker your face into wrinkles ten years ahead of time, by a self-inflicted fit of the dismals.

I went into the country with a determination to succeed. As others had there succeeded, I could not be induced to believe that failure in so simple an enterprise could overtake me, as I felt myself quite as competent as they. A resolute will overcomes all difficulties. It was one of the leading characteristics of Napoleon to regard nothing as impossible. His astonishing successes are to be attributed to his indomitable will, scarcely less than to his vast military genius. Wellington was distinguished for a similar peculiarity. The entire Peninsular campaign was, indeed, but one long display of an iron will, resolute to conquer difficulties by wearing them out. Alexander the Great was quite as striking an example of what a powerful will can effect. His stubborn determination to subdue the Persians; his perseverance in the crisis of battle, and the emulation to which he thus stimulated his officers and men, did more for his wonderful career of victory than even his great strategic abilities. In the life and death struggle between England and France, during the first fifteen years of this century, it was the stubborn will of the former which carried the day; for though Napoleon defeated the British coalitions again and again, yet new ones were as constantly formed, until at last the French people, if not their emperor, were completely worn out. The battle of Waterloo, which was the climax of this tremendous struggle, was also an illustration of the sustained energy, the superior will of the British. In that awful struggle, French impetuosity proved too weak for English resolution. "We will see who can pound the longest," said Wellington; and as the British did, they won the battle.

It is not only in military chieftains that a strong will is a jewel of great price. Nations and individuals experience the advantages of a resolute will; and this alike in large and small undertakings. It was the determined will of our forefathers to which, with divine help, we are principally indebted for our freedom. For the first few years after the declaration of independence, we lost most of the battles that were fought. New York and Philadelphia were successively captured by the foe; South Carolina fell; New Jersey was practically reannexed to England; almost every thing went against us. Had the American people been feeble and hesitating, all would have been lost. But they resolved to conquer or die. Though their cities were taken, their fields ravaged, and their captured soldiers incarcerated in hideous prison-ships, they still maintained the struggle, making the pilgrimage of freedom with naked feet, that bled at every step. Had our fathers been incapable of Valley Forge, had they shrunk from the storm-beaten march on Trenton, we should never have been an independent nation. There are people in the Old World, full of genius and enthusiasm for liberty, who yet cannot achieve freedom, principally, perhaps, because they lack the indomitable will to walk the bloody pilgrimage. The outbreak of the slaveholders' rebellion covered the Union armies with defeat at numerous points, because rebellions are always successful at the beginning. But the determined will to crush out treason will eventually overwhelm and master it.

A strong will is as necessary to the individual as to the nation. Even intellect is secondary in importance to will. A vacillating man, no matter what his abilities, is invariably pushed aside, in the race of life, by the man of determination. It is he who resolves to succeed, who begins resolutely again at every fresh rebuff, that reaches the goal. The shores of fortune are covered with the stranded wrecks of business men who have wasted energy, and therefore courage and faith, and have perished in sight of more resolute but less capable adventurers, who succeeded in making port. In fact, talent without will is like steam dissipating itself in ' the atmosphere; while abilities controlled by energy are the same steam brought under subjection as a motive power. Or will is the rudder that steers the ship, which, whether a fast-sailing clipper or a slow riverbarge, is worthless without it. Talent, again, is but the sail; will is what drives it. The man without a will is the puppet and bubble of others by turns. The man with a will is the one that pulls the strings and catches the dupes. Young man, starting out in life, have a will of your own. If you do not, you will drag along, the victim of perpetual embarrassment, only to end in utter ruin. If you do, you will succeed, even though your abilities be moderate.

All this may be viewed as a digression. But it is not so. I do not write for the rich and prosperous, but for those who have been unsuccessful. They need encouragement and bracing up. If their experience has been disastrous, that of others, who have succeeded, should be set before them. Some fifty years ago there lived in this city an old man, who by dint of tact, with the aid of keen perceptive faculties, had acquired much celebrity with a large class of his neighbors as something between a prophet and a fortune-teller. He did not, however, assume the character either of a religious fanatic or of a crafty disciple of Faustus. But he was well read in the Scriptures; he had a good share of common sense, and a voluble tongue, and by degrees he attained a fame for wise sayings and for capability to advise, which he owed more to his natural talents and a loquacious disposition than to any less worthy means. Being advanced in years, and his lot humble, he turned the good opinion formed of him to the account of his livelihood, by discussing questions put to him by his visitors in a frank and manly spirit; and without ever demanding recompense, he was ready to receive any gratuity that was offered by them on their departure. Moreover, his advice was always, if not valuable, at least good in kind; and few quitted his humble dwelling without leaving their good wishes in a substantial shape, or without having also formed a favorable opinion of their mentor.

At length, so extensive did this good man's fame become, that many from curiosity alone were induced to visit him, and hear his wise sayings. His counsel was usually couched in short and terse sentences; frequently in proverbs, and often in the language of the Bible, to which he would sometimes refer his inquirers for passages which he said would be found applicable to their case. As these passages were usually selected from the Proverbs, and other books of somewhat similar description, which contained some rule of morals, or which advocated the Christian duties, he seldom failed to be right. Among others who were led by curiosity to this wise man, was a young farmer, then not long entered upon the threshold of life, whom, after some of the Scripture references adverted to, he dismissed with the parting advice, "To keep a smiling countenance, and a good exertion." The young farmer lived to become an old man, and is now gathered to his fathers. But for many years I heard him from time to time revert with pleasure to his visit, and say that this simple aphorism had frequently cheered him in the hour of difficulty; and that the thought of the old man's contented countenance and encouraging voice when he uttered it, had gone far to make him place confidence in his counsel.

We are all too prone to brood over the clouds upon our atmosphere, and too feebly do we keep the éye of hope fixed on the first sunbeam which breaks through as the symbol of their dispersion. In reality, most of them are merely passing clouds. Some glances at a blacker picture still, will go far to clothe with brighter hues the less gloomy picture which may happen to be our own. Thus, with " a smiling countenance and a good exertion," let every one of us, whether his lot be cast with the plough, the loom, or the anvil, put forth manfully his powers, and, thankful to a gracious God for the blessings yet spared, be it our effort in our worldly duties to follow the example set us in higher things, " forgetting those things that are behind, and reaching forth unto those which are before, let us press towards the mark for the prize and if we thus demean ourselves, we shall not fail, in earthly any more than in spiritual things, to obtain our reward.

All know that one effect of the rebellion was to paralyze nearly every kind of business, suddenly enriching the few, but as suddenly impoverishing the many. On my quiet little plantation I was entirely beyond the reach of its disastrous influence. It lost me no money, because my savings had been loaned on mortgage. It is true that interest was not paid up as punctually as aforetime, but the omission to pay occasioned me no distress; hence I occasioned none by compulsory collection. The summer of 1861, however, did reduce prices of most of my productions. The masses had less money to spend, and therefore consumed less. Yet my early consignments of blackberries sold for twenty-five cents a quart, and the whole crop averaged fourteen. My strawberries yielded abundantly, escaping the frost which nipped the first bloom of all other growers, no doubt protected by the well-grown peach-trees, and netted me sixteen cents. Raspberries bore generously, and netted quite as much; while peaches, though few in number, brought the highest prices. The total income that year was certainly less than usual, by several hundred dollars – but what of that? It was double what I needed to support my family. Thus, no national disaster, no matter how tremendous, seems able to impoverish the farmer who is free from debt. Nothing short of the tramp of hostile armies over his green fields can impoverish such a man.

CHAPTER XXV.
WHERE TO LOCATE – EAST OR WEST.

EVERY great national calamity has the effect of driving men from the cities to engage in agriculture. Such has been the result of the late Avar for the Union. I have been in a position to observe its operation on the minds of hundreds whom it covered with disaster. There has been the usual desire to break away from the cities, and settle in the country. The life-long convictions of my own mind have taken possession of the minds of others. Property in the cities ceased, for a time, to be salable, while farms have been in more general demand than for years past. Foreign immigration was measurably stopped, because men fly from convulsions, not to countries where they are to be encountered. When war desolates the nations of Europe, the people migrate hither to avoid its horrors; when it desolates ours, they remain at home.

During the late disastrous experiences of city life, many of my friends upon whom they fell with great severity were free in their congratulations on my happy change of life. They had been as free in doubting the propriety of my experiment. Now, however, they looked up to me as possessing superior sagacity; were desirous themselves of imitating my example, and sought instruction and advice as to how they should proceed. Three of them are already located near me; so that, instead of cutting entirely loose from old associates by coming into the country, I have attracted them into a closer intimacy than ever. Dear as my home was without them, it is rendered doubly dearer by close association with long-tried friends.

Location is perhaps the most, important consideration. A cash market all the year round for every variety of produce that a man can raise, is of the utmost importance to secure. Such is invariably to be found in close proximity to the great cities; and there, singularly enough, the wealthiest farmers in the Union will generally be found.

When we go to the extreme North, where their market is limited, and where they produce only the heavy grains, and grasses, farming is so little an object that improved places can always be bought for less than their cost. It is very frequently the same throughout the West, where so much that is raised upon a farm is valueless; and where, for even the grains, they have a market which barely pays the expense of living. The expense incurred in farming can be regulated by the profit of the crops; and where even no manure is required, the labor has to be expended, and crops in distant localities often fail to pay the expense of this labor. Where land will pay for a liberal cultivation, as well as fertilizing, it is much better, as a farmer must work his stock, and a certain amount of care is indispensable. The difference in value existing between those farms near a market and those remote from it, is enormous. If the mind will consider the immense amount of produce in the way of fruits and vegetables which, near a city, will command the highest prices, and which at a distance are an entire loss, a conception can be readily formed of what they amount to in dollars and cents.

Land in Illinois and Iowa can be purchased for a dollar an acre, but corn is at times of so little value as to be consumed for fuel. The wheat crop is annually decreasing in its acreable product, because no one values or applies manure. The West may be the paradise of the European immigrant, who, having abandoned friends and home, may with propriety settle in one spot as well as in another; because, go where he will, he will be sure to find none but strangers. But for residents of our cities who go thither, very few acquire property by legitimate farming, even after sacrificing all the tender associations of relatives and friends whom they leave behind, and enduring hardships and trials of double severity with those they need encounter if they would consent to suffer them on lands within thirty miles of their birthplace. If they become rich, it is by hazardous speculation, or by the rise in value of their lands. So far as real, practical farming is concerned, it will be found that the East is incomparably superior to the West; but, so far as small farmers like myself are concerned, it would be folly to deny this superiority.

I say nothing as to the superior ease with which corn and wheat are produced in the two sections, but refer only to the amount of money that can be realized from an acre there and an acre here. Beyond question, there are certain crops that are produced with greater ease in the West than in the East; but of what value is this superior facility if it does not pay? I have cleared from a single acre of tomatoes more than enough to buy a hundred and sixty acres in Iowa. If I had located there, who would have been ready to buy my abundant crop of berries? The truth is, that it is population that gives value to land, – population either on it or around it, – to convert it into lots covered with buildings, or to consume whatever it may produce. The West is a glorious region for the foreign immigrant, or for him who was born upon the rugged hill-sides of the Eastern States, but it is not the proper location for the class for whose instruction these pages have been written.

Few persons who have been nurtured and educated all their days in Eastern cities, and who have probably never been more than fifty miles from home, have any correct idea of what this gigantic West really is until they reach the spot itself. Why leave the privileges of a long-established civilization, – the schools, the churches of home, – the daily intercourse of acquaintances and friends, – merely because land producing twenty bushels of wheat per acre can be purchased for a dollar, when that producing twenty times as much in fruit or vegetables can be had for fifty, and often even for less? I doubt not there must be many in that region who now wish themselves back in their old homes.

If my example be worth imitating, land should be obtained within cheap and daily access to any one of the great cities. If within reach of two, as mine is, all the better, as the location thus secures the choice of two markets. In Pennsylvania, all the land around Philadelphia is held at high prices. Much of it is divided into small holdings, many of which are rented to market gardeners at prices so high that none but market gardeners can afford to pay them. Others are worked by their owners, who live well by feeding the great city. Gradually, as the city extends in every direction, these small holdings are given up to streets and buildings, thus enriching their owners by the rise in value. The truckers move further back, where land is cheaper. But the modern facilities for reaching the city by railroad have so greatly multiplied, that they are practically as near to it as they were before. The yield from some of these small holdings is very large. But the cost of land thus situated was too great for my slender capital when I began.

Hence I sought a location in New Jersey. There unimproved land, within an hour of Philadelphia, can be purchased for the same money per acre which is paid in Pennsylvania as annual rent. For ten to twenty dollars more, in clearing up and improving, it can be made immediately productive, as the soil of even this cheap land is far more fertile than is generally supposed. Thousands of acres of this description are always for sale, and thousands are annually being bought and improved, as railroads and turnpikes leading to the city are being established. Many Germans have abandoned the West, and opened farms on this cheap and admirably located land, from which they raise prodigious quantities of fruit and truck for Philadelphia and New York.

Colonies of New Englanders, allured by the early season, as compared with that of their own. homes, the productive soil and the ready access to market, have settled upon and around the new railroad just opened, which leads south from Camden through the town of Malaga, where a large tract has recently been divided into farms of various sizes. They bring with them all the surroundings of an advanced civilization.

To those with no capital but their own labor and a determination to conquer success, these lands offer the highest inducements. Most of them can be had on credit, by men who will settle and improve, at twenty to thirty dollars per acre, within a little over an hour's ride to Philadelphia. This tract is distant but a few miles from the Delaware river, and probably no better could be found. Any number of locations can be had. Many are already improved by buildings, fencing, and all the preliminary comforts which cluster round an established home. The settler may choose between the improved and the unimproved.

But there is a better country north of Camden, lining the shore of the Delaware, where any number of locations may be found, improved by buildings, and at moderate prices, as well as on favorable terms as to payment. Vast progress in improvement has been made through all this region within ten years. New towns have been built, new turnpikes constructed, while the great railroad puts the cultivator in constant connection with the two overgrown cities at its termini. Land is increasing in value as population flows in. The margin of the Delaware, from Philadelphia upward, is being lined with villages, between which new farm-houses and cottages are annually erected; and the young of this generation will live to see it a .continuous settlement of substantial villas, peopled by the swarms of educated families which a great human hive like Philadelphia is annually throwing off. A location within such an atmosphere of improvement must continually increase in value. The owner will find himself growing richer from this cause, just as the trucker on the Pennsylvania side has done – not so rapidly, but quite as surely. An investment in such land, properly managed, and not permitted to deteriorate, will assuredly pay. My own little farm is an illustration; for more than once have I been solicited to sell at double the price it cost me.

I am now looking at the future, as well as at the present. Yet the apparent anomaly of there being always an abundance of land for sale in so desirable a district, must not be overlooked. But it is so throughout our country; there are always and everywhere more sellers than buyers. It is the same thing in the cities; everywhere there is somebody anxious to sell. It would seem that we either have too much land in this country, or too small a population. Time alone can produce the proper equilibrium. The land cannot be increased in quantity, but it is evident that the population will be. As this is not a treatise either upon land or farming, but the experience of a single individual, so each claimant for a similar experience must choose for himself.

But choose as he may, locate as he will, he must not, as he hopes to succeed in growing the smaller fruits to profit, locate himself out of reach of a daily cash market. New York and Philadelphia may be

JAMES MILLER

likened to two huge bags of gold, always filled, and ever standing open for him to thrust in his hand, provided in the other he brings something to eat. From this exhaustless fountain of wealth, whole adjacent populations have become rich. The appetite of the cities for horticultural luxuries has revolutionized the neighboring agriculture, enhanced the value of thousands of acres, infused a higher spirit into cultivators, elevated fruit-growing into a science, and started competition in a long rivalry after the best of every thing that the earth can be made to yield. All this is no spasmodic movement. It will go on for all future time; . but in this grand and humanizing march after perfection in producing food for man, the careful tiller of the soil, with moderate views and thankful heart, will be sure to find TEN ACRES ENOUGH.

THE END.

www.ingramcontent.com/pod-product-compliance
Lightning Source LLC
Chambersburg PA
CBHW051913050825
30591CB00042B/808

* 9 7 8 1 6 6 7 3 0 6 3 7 7 *